Lecture Notes in Economics and Mathematical Systems

Managing Editors: M. Beckmann and W. Krelle

357

Peter Zörnig

Degeneracy Graphs and Simplex Cycling

Springer-Verlag
Berlin Heidelberg New York London Paris
Tokyo Hong Kong Barcelona Budapest

Author

Peter Zörnig
Universität Bochum
Sprachwissenschaftliches Institut
Postfach 10 21 48, W-4630 Bochum 1

ISBN-13: 978-3-540-54593-4   e-SBN-13: 978-3-642-45702-9
DOI: 10.1007/978-3-642-45702-9

Typesetting: Camera-ready by author

42/3140-543210 – Printed on acid-free paper

Foreword

The first monograph on "Degeneracy Problems" appeared 1986 as "Degeneracy Graphs and the Neighbourhood Problem" by H.-J. Kruse as No 260 of Lecture Notes. The contents of that book was based upon a simple problem of degeneracy in linearly constraint optimization problems which has been posed in 1976. Since then the literature concerning degeneracy grew and a part of the corresponding research throughout the world has been devoted to the theory and application of degeneracy graphs. It turned out that the better should be the use of degeneracy graphs to solve practical problems of various kind the more theoretical knowledge about such graphs is needed. Just to mention one of such problems, let us remind of the cycling of the simplex method. This phenomenon is known since the beginning of the 50-ies and caused some trouble. Several "anticycling" methods have been elaborated as the time passed and even in the late 80-ies new proposals have been published. Strange enough, there has been no publication in which a trial was made to explain this phenomenon, i.e. to pose the question "When or under which conditions simplex-cycling occurs?".

The author of this book makes a valuable contribution to the theory of the degeneracy graphs and answers the above question.
Based on his results the research into degeneracy and degeneracy graphs can now turn to new unsolved degeneracy problems, which would remain unsolvable without the knowledge presented in this book.

Hagen, May 1991 Tomas Gal

Acknowledgements

The present publication is based on my doctoral thesis which was submitted to the Fernuniversität Hagen (FRG). I wish to express my sincere thanks to my advisor Prof. Dr. Dr. T. Gal for his encouragement and advice. I am also grateful to Prof. Dr. G. Fandel, who supported the writing of this book.

I am indebted to my former collegues and friends Dipl.-Math. M. Buhlmann, Dipl.-Math. F. Geue, Dr. H.-J. Kruse and Dr. G. Piehler for useful suggestions.

I would also like to thank Prof. Dr. E. Hopkins for corrections and valuable improvements on my earlier English translations.

The layout and typography were carried out by U. Roos and A. Hennern, who overcame all technical difficulties in their careful preparation of the manuscript. Some of the figures were drawn by Dipl.-Ing. U. Rothe.

Last but not least I am grateful to Prof. Dr. G. Altmann, who enabled me to write this book while I was engaged in his project "Language Synergetics".

Bochum, May 1991 Peter Zörnig

To my parents
who could not live to see
the publication of this book

Contents

Index of symbols

$A, \bar{A}$	(enlarged) matrix of coefficients
b	right-hand side vector
$B, B_{j_1,\ldots,j_\sigma}$	basis (with index set $\{j_1,\ldots,j_\sigma\}$)
B^0	basis set of the vertex x^0
$B \leftrightarrow B^*$	possibility to pass
$B \overset{+}{\leftrightarrow} B^*$	from B to B^* in one pivot step
$B \overset{-}{\leftrightarrow} B^*$	with an (arbitrary, positive, negative) pivot element
β_i	cf. Section 4.3.1
$c, \bar{c}$	(enlarged) vector of objective function coefficients
c_B	vector of objective function coefficients assigned to basis B
C	cycle of a degeneracy graph or simplex cycle
$d, \bar{d}$	(enlarged) vector of objective function coefficients belonging to the reduced linear optimization problem
$d(v, w)$	distance between the nodes v and w of a graph
$d(G)$	diameter of the graph G
$\tilde{d}$	cf. Section 4.3.1
$D_{j_1,\ldots,j_\sigma}$	subdeterminants of the initial tableau belong-
$\bar{D}_{j_1,\ldots,j_\sigma,\nu}$	ing to a reduced linear optimization problem
$\bar{D}^i_{j,k}, D^i_j$	(cf. Sections 4.4.1 and 5.2.5.1)
$\mathcal{D}(\mathcal{S})$	σ-normal representation system of $\mathcal{S}$
Δz_j	relative cost coefficients
δ	minimum degree of a graph
e_u	u-th unit vector
E	edge set of a graph
$E_\nu, E_{a,b}$	(constraint-)hyperplane
$g(v)$	degree of the node v
$G_1 \cong G_2$	G_1 is isomorphic to G_2
$G(X)$	representation graph of X
$G'(X)$	graph of the polytope X
G^0, G^0_+, G^0_-	general, positive and negative degeneracy graph of x^0
G_Y, G^+_Y, G^-_Y	general, positive and negative canonical degeneracy graph of x^0

$G_{Y,d}, G^+_{Y,d}, G^-_{Y,d}$	general, positive and negative LP-degeneracy graph
$G^{\sigma,n}$	complete $\sigma \times n$-index graph
$H_\nu, H_{a,b}$	halfspace of $\mathbb{R}^n$
int X	interior of X
I	index set
I_m	$m \times m$-unit matrix
j^+, j^+_ν	index of the entering basic variable
K	edge of a graph
K_C	cone induced by C
$(K_1, \ldots, K_n)$	edge path from K_1 to K_n
$K(p_1, \ldots, p_r)$	complete r-partite graph
$L(G)$	line graph of G
$\mathbb{N}$	set of natural numbers
ω, ω'	node-/edge-connectivity
$p(n)$	number of (unordered) partitions of n
P_C	point set induced by C
$\mathbb{R}, \mathbb{R}^n$	set of real numbers/ vectors with n real components
$\mathbb{R}^{m \times n}$	set of $m \times n$-matrices with real elements
$\mathcal{S} = \{s_1, \ldots, s_p\}$	σ-homogeneous set system
$\bar{\mathcal{S}}$	complementary system of $\mathcal{S}$
$< \mathcal{S} >$	subgraph of $G^{\sigma,n}$ induced by $\mathcal{S}$
S_C	star-shaped graph
σ	degeneracy degree
$T, T_{j_1,\ldots,j_\sigma}$	tableau (corresponding to basis $B_{j_1,\ldots,j_\sigma}$)
$\mathcal{T} = \{t_1, \ldots, t_q\}$	(σ-normal) representation system
$\mathcal{T}_Y$	set system induced by Y
u	vector of slack variables
U	number of nodes of a graph
U'	number of edges of a graph
$U(x^1, \ldots, x^k)$	vector space generated by $x^1, \ldots, x^k$
$U(x^0; x^1, \ldots, x^k)$	affine subspace parallel to $U(x^1, \ldots, x^k)$
v	node of a graph
$(v_1, \ldots, v_n)$	trail from v_1 to v_n
V	node set of a graph
V^*	node set of $L(G)$
$V^*_{i,j}$	component of a partition of V^*

$V(x,y)$	closed segment between x and y
W	path of a graph
$x, \bar{x}$	(enlarged) vector of variables
$x_B, x_B^{(0)}$	(complete) basic solution of the basis B
x^0	vertex of X
$X, \bar{X}$	solution set of a linear optimization problem
y^j	column of $\bar{Y}$
$\tilde{y}^j$	cf. Section 4.3.1
$Y, \bar{Y}$	(enlarged) matrix of coefficients of a reduced linear optimization problem
z	objective function value

1. Introduction

Problems with linear constraints are of special importance in the topic of mathematical optimization. The feasible solution set X represents a convex polyhedral set. In practice the case that X contains degenerate (overdetermined) vertices, occurs very often[1]. Degeneracy involves many different problems, not only in connection with the determination of an optimal solution by a "vertex searching" procedure but also in postoptimal analysis (cf. Chapter 2). Shortly after Dantzig (1951) published the simplex method, a number of papers on the degeneracy phenomenon appeared[2]. Recent articles[3] demonstrate that many questions are still open[4]. The reason for all kinds of degeneracy problems is the fact that a great number of bases is associated with a degenerate vertex[5]. The extensive literature shows that degeneracy problems have always been treated separately from each other. The idea of studying all these problems from a common point of view led to the investigation of so-called degeneracy graphs (cf. Section 3.1). These graphs represent appropriate tools for describing the complex structure of the basis or tableau set associated with a degenerate vertex and thus for revealing the theoretical background of the emergence of different degeneracy phenomena. Initial investigations of these graphs were carried out by Gal (1978, 1985) in connection with the problem of determining all neighbours of a degenerate vertex. Degeneracy graphs also render valuable help in the solution of different other kinds of degeneracy problems. Based on the theory of degeneracy graphs it has already been possible to develop procedures to solve the so-called neighbourhood

---

[1] Cf. e.g. Greenberg (1986:636, 650), Perold(1980:240) and Section 3.1.1.

[2] Cf. e.g. Charnes (1952), Hoffmann (1953), Nelson (1957).

[3] Cf. Balinski et al. (1986), Cameron (1987), Cirina' (1985,1989), Clausen (1987), Fourer (1988), Greenberg (1986), Hattersley/Wilson (1988), Horst et al. (1988), Megiddo (1986), Mlynarovic (1988), Nygreen (1987), Ramesh et al. (1987), Ryan/Osborne (1988), Sherali/Dickey (1986), Vörös (1987).

[4] Perhaps the "competing procedures" of Khachian (1979) and Karmarkar (1984) have motivated scientists to reflect on the "classical" simplex method and different aspects of degeneracy connected with it.

[5] Cf. Dantzig (1966:Ch. 10), Gal (1985), Kruse (1986: Ch.4) among others.

problem and to achieve better understanding of the theoretical foundation of sensitivity analysis and the determination of shadow prices under degeneracy. For the solution of the neighbourhood problem the fact that every positive degeneracy graph contains an *N*-tree[6] is fundamental; for sensitivity analysis and the determination of shadow prices the question whether optimum graphs[7] are connected is decisive. Surveys of theory and application of degeneracy graphs are given in Gal (1988) and Gal/Kruse/Zörnig (1986,1988).

The present publication starts from the results above. It has essentially two aims:

1) In order to make the use of degeneracy graphs more effective, it is desirable to know as much as possible about their structural properties. Therefore the theory of degeneracy graphs will be developed generally.
2) Degeneracy graphs will be used to explain a special degeneracy problem, namely the simplex cycling in linear optimization. The intention here is not to develop a further anticycling rule, but to investigate the following unsolved question: "Under which conditions does the phenomenon of simplex cycling occur at all?" The answer to this question is of theoretical as well as practical importance.

The present book is divided into five chapters. Below a summary of different degeneracy problems in the topic of mathematical optimization is to be found. In Chapter 3 the theory of degeneracy graphs is developed. A characterization of these graphs is derived which especially implies certain structural properties (numbers of nodes and edges, connectivity etc.). Applications of these results are also mentioned (Section 3.2.3). The last two chapters deal with the problem of simplex cycling. Chapter 4 presents diverse concepts to explain simplex cycling using degeneracy graphs. For each case necessary and sufficient conditions for the occurrence of simplex cycling are derived. Among other things it will become evident that simplex cycling occurs

---

[6] Cf. e.g. Kruse (1984,1986) and Gal/Kruse (1984).

[7] Cf. Kruse (1987), Piehler (1988), Piehler/Kruse (1989) and Section 2.3.

if and only if the positive degeneracy graph can be enlarged to a "star-shaped" graph contained in the LP-degeneracy graph. Based on these results Chapter 5 evolves procedures to construct cycling examples. They permit carrying out certain tests to clarify different practically oriented questions concerning simplex cycling. For example the efficiency of anticycling rules can be tested by applying them to a set of constructed cycling examples[8].

2. DEGENERACY PROBLEMS IN MATHEMATICAL OPTIMIZATION

Simplex cycling in linear optimization is the most popular degeneracy problem[9] (cf. Ch. 4). Moreover, the degeneracy phenomenon causes problems in many other fields of mathematical optimization with regard to convergence and the efficiency of the algorithms determining the optimal solution. The following selection[10] of problems will provide an impression of the extension of the topic "degeneracy" and its practical importance.

---

[8] Further applications of constructed cycling examples are mentioned in Chapter 5.1.

[9] This problem was already investigated in the fifties (cf. Beale (1955), Charnes (1952), Dantzig/Orden/Wolfe (1955), Hoffmann (1953)).

[10] Numerous references are adduced from Mathies (1989). Further fields in which degeneracy causes difficulties are: Neighbourhood problem (Gal (1985) and Kruse (1984,1986)), methods for solving systems of linear inequalities or for determining the vertices of a convex polyhedral set (Dyer/Proll (1980,1982), Sherali/Dickey (1986), Wallace (1985)), parametric programming (Gal (1979), Ritter (1984)), vector maximization (Philip (1977), Proll (1987)), portfolio analysis (Vörös (1987)), infinite-dimensional linear optimization (Nash (1985)), piecewise-linear optimization (Fourer (1988), Ruszczynski (1986)), global optimization (Horst et al.(1988)), fixed cost transportation (Ahrens/Finke (1975), Mc Keown (1978)), linear minimax optimization (Ahuja (1985)), linearly constrained optimization (Calmai/More'(1987, Section 4)), linear optimization with variable upper bounds ((Todd (1982)), reduced linear optimization (Tomlin/ Welch (1983)).

2.1 CONVERGENCE PROBLEMS IN THE CASE OF DEGENERACY

2.1.1 CYCLING IN LINEAR COMPLEMENTARITY PROBLEMS

Linear complementarity problems occur in

a) the solution of quadratic optimization problems,
b) bimatrix games[11].

To a: In order to solve a *quadratic optimization problem* of the form[12]

$$\left.\begin{array}{ll} \min z = & \frac{1}{2}x^TQx + q^Tx \\ \text{s.t.} & \\ & Ax \leq b \\ & x \geq 0, \end{array}\right\} \qquad (2.1.1.1)$$

where $Q \in I\!R^{n\times n}$ is a symmetric, positive definite matrix and $A \in I\!R^{m\times n}, q, x \in I\!R^n, b \in I\!R^m$, the Kuhn-Tucker conditions are generally used. They are necessary and sufficient for an optimal solution of (2.1.1.1) and can be represented in the form

$$\left.\begin{array}{ll} w - Mz & = r \\ w, z & \geq 0 \\ w^Tz & = 0 \end{array}\right\} \qquad (2.1.1.2)$$

where $M \in I\!R^{p\times p}, w, z, r \in I\!R^p$. The variables w, z must be determined such that (2.1.1.2) is fulfilled (M, r fixed). The system (2.1.1.2) is called a *linear complementarity problem.* The solution of (2.1.1.1) can be obtained immediately from the solution of (2.1.1.2).

To b: A *Bimatrix game* (i.e. a nonzero-sum two person game) can also be formulated as a linear complementarity problem (2.1.1.2). The solution represents the equilibrium point[13].

---

[11] Cf. Lemke (1965:681).

[12] Cf. Horst (1987:270).

[13] Cf. Bitz (1988:23ff.), Schwödiauer (1987:11ff.).

In order to solve (2.1.1.2) the algorithm of Lemke (1965) is generally used (cf. also Horst (1987:377ff.)). This procedure starts with a tableau of the form

$$w|z_0| - M|r \tag{2.1.1.3}$$

and solves (2.1.1.2) by appropriate pivot steps. Kostreva (1979) detected that Lemke's algorithm cycles for certain constructed degenerate problems. Cycling is also possible when Keller's (1973) algorithm is applied (cf. Chang/Cottle (1980)) which is a pivoting method starting with a "symmetric" modification of (2.1.1.3) to solve (2.1.1.2). However, the use of Bland's (1977) anticycling rule guarantees the convergence of Keller's algorithm.[14] .

2.1.2 CYCLING IN NETWORK PROBLEMS

A *network problem* or *transshipment problem* is a linear optimization problem of the form (cf. Bixby/Cunningham (1980) a.o.):

$$\left.\begin{array}{lll} \min\ z = & c^T x & \\ \text{s.t.} & & \\ & Nx & = b \\ & x & \geq 0, \end{array}\right\} \tag{2.1.2.1}$$

where $c, x \in \mathbb{R}^n, b \in \mathbb{R}^m$ are vectors and $N \in \mathbb{R}^{m \times n}$ is a so-called node-arc incidence matrix, the elements being +1,-1 or 0 (cf.also Solow (1984:348)). *Transportation* and *assignment problems* are special cases of (2.1.2.1) which play an important role in practice. The former are generally solved by procedures based on the vertex searching method (MODI method, Stepping-Stone algorithm) developed by Dantzig (1951a) and Charnes/Cooper (1954)[15]. Since transportation problems

---

[14] Cf. Chang/Cottle (1980:128ff.).

[15] The only difference between these procedures consists in the computation of opportunity costs (cf.Domschke (1981:102), Ohse (1987:267)).

in practice are often "highly degenerate"[16] , cycling can occur using the above methods. Gassner (1964) pointed out that the MODI method cycles, when it is applied to specially constructed $n \times n$-assignment problems[17] with $n \geq 4$. Subsequently Madan Lal Mittal (1967:176ff.) developed a modification of the MODI method that insures the convergence. A finite variant of the simplex method for solving network problems which uses "strongly feasible trees" was elaborated by Cunningham (1976). Especially in solving transportation problems, this procedure coincides with the method of Barr/Glover/Klingman (1977,1978). A general cycling example for the network simplex method was introduced by Cunningham (1979)[18].

2.1.3 CYCLING IN BOTTLENECK LINEAR PROGRAMMING

A *bottleneck linear programming problem* is of the form

$$\left.\begin{array}{rl} \min z = & \max\{c_j | j \in \{1,\dots,n\}, x_j > 0\} \\ \text{s.t.} & \\ & Ax = b \\ & x \geq 0, \end{array}\right\} \qquad (2.1.3.1)$$

with $c = (c_1,\dots,c_n)^T, x = (x_1,\dots,x_n)^T \in I\!R^n, A \in I\!R^{m \times n}, b \in I\!R^m$ (cf. Frieze (1975:871)). In analogy to network problems, model (2.1.3.1) includes the *bottleneck transportation problem* and the *bottleneck assignment problem* as special cases. Different practical problems can be represented by (2.1.3.1), e.g. the problem of parallel transportation of perishable goods, which have to reach their destinations as fast as possible (cf. Garfinkel/Rao (1976:291)).

---

16 Cf. Boenchendorf (1987:87), Madan Lal Mittal (1967:175).

17 Note that the assignment problem is the most degenerate form of the transportation problem (cf. Akgül (1987)).

18 Cunningham/Klincewicz (1983) showed that any cycling example for the network simplex method has a length of at least ten.

In order to solve problem (2.1.3.1), simplex type algorithms have been developed[19]. Though no cycling example for bottleneck linear programming is mentioned explicitly in literature, the possibility of cycling can not be excluded. Seshan/Achary (1982:349) propose the application of the "lexicographic rule" of Dantzig/Orden/Wolfe (1955) to insure the convergence.

2.1.4 CYCLING IN INTEGER PROGRAMMING

A pure *integer programming problem* is of the form

$$\left.\begin{array}{lll} \max z = & c^T x & \\ \text{s.t.} & & \\ & Ax & \leq b \\ & x & \geq 0 \\ & x & \text{integer vector} \end{array}\right\} \qquad (2.1.4.1)$$

where $x, c \in \mathbb{R}^n, A \in \mathbb{R}^{m \times n}, b \in \mathbb{R}^m$, and all elements of c, A, b are also integers. The so-called cutting plane algorithms (cf. e.g. Gomory (1963), Young (1968)) represent an important class of solution methods for (2.1.4.1). The common principle of these procedures is the following[20]: First of all problem (2.1.4.1) is solved without the integer requirements. If the solution is not an integer vector, another cutting plane is added to system (2.1.4.1) which "cuts off" the just found solution without changing the set of feasible solutions. Now the new problem has to be solved without integer requirements, etc. In (primal) cutting plane algorithms the case of degeneracy (i.e. the right-hand side of the "cutting constraint" equals 0; cf. Burkard (1972:237ff.)) involves specific difficulties which anticycling rules can not eliminate[21]. In general linear optimization an anticycling rule insures the termination of the procedure, since

[19] Different solution methods are surveyed in Seshan/Achary (1982:347).

[20] Cf. Burkard (1987:379ff.).

[21] Degeneracy in integer programming not only involves the danger of cycling but also reduces the efficiency of solution methods (cf. Balas (1971), Glover (1968:729), Nygreen (1987), Tomlin (1970)).

a) only a restricted number of feasible basic solutions associated with vertices exists

and

b) the anticycling rule avoids the repetition of basic solutions.

In cutting plane algorithms it is generally impossible to determine upper bounds for the number of cutting planes or the number of vertices of the solution set. Hence assumption a) is not fulfilled, i.e. the termination is not guaranteed even if the repetition of basic solutions is impossible. Young (1968) developed appropriate modifications to insure the termination of the primal cutting plane algorithm in the case of degeneracy.

2.2 EFFICIENCY PROBLEMS IN THE CASE OF DEGENERACY

2.2.1 EFFICIENCY LOSS BY WEAK REDUNDANCY

A constraint of a linear optimization problem is called redundant if it can be omitted without effecting the set of feasible solutions (cf. Karwan et al.(1983:1)). If a redundant constraint is active for any vertex of the solution set, it is called weakly redundant[22]. In this case the corresponding vertex is degenerate. Thus weak redundancy is a special case of degeneracy occuring especially if the system of constraints consists of different subsystems which have to be preserved individually[23] or if the solution method requires the addition of further constraints (e.g. cutting plane methods in integer programming)[24]. The loss in efficiency caused by (weak) redundancy may be considerable. Since redundant constraints have to be transformed in each iteration, they

[22] Cf. Gal (1983,1987:160), Kruse (1984:12), Nelson (1957:405f) and Section 3.1.1.

[23] Cf. Tomlin/Welch (1983:233f).

[24] Cf. Müller-Merbach (1973:386f).

require additional computations and storage capacity. Though redundancy may be advantageous (it enables the application of special solution procedures in some cases), its negative aspects are generally speaking predominant.[25] Therefore diverse procedures for detecting redundant constraints have been evolved[26]. Omitting redundant constraints, however, causes difficulties with regard to the economical interpretation of the optimal solution[27]; e.g. the determination of shadow prices or the determination of critical intervals in sensitivity analysis occasionally leads to incorrect results when redundant constraints have been omitted[28].

2.2.2 EFFICIENCY PROBLEMS FROM THE PERSPECTIVE OF THE THEORY OF COMPUTATIONAL COMPLEXITY

The theory of computational complexity[29] provides the necessary concepts for estimating the efficiency of algorithms. The maximal or average complexity is considered as a function of the problem size, characterized by appropriate parameters (e.g. by the numbers m,n of constraints or variables in linear optimization). Numerous investigations have already been carried out to determine the efficiency of the simplex algorithm or any of its variants.

a) Worst-case analysis

Dantzig's (1951) simplex method proved to be good in practice over several years[30]. But in accordence with Edmonds (1965)[31] an algorithm

---

[25] Cf. Karwan et al.(1983:6).

[26] Cf. Bixby/Wagner (1987), Karwan et al.(1983), Telgen (1983) among others.

[27] Cf. Zimmermann/Gal (1975).

[28] Cf. Gal (1988, Section 5.4) and Section 2.3.

[29] Cf. Bachem (1980).

[30] Cf. Borgwardt (1987:15), Gale (1969), Gotterbarm (1983:573), Liebling (1973:248), Smale (1983:241f).

[31] Cf. also Derigs (1986:48).

is considered as "good" only if the maximal computational complexity is a polynomial of the problem size[32]. Klee/Minty (1972) showed that the simplex algorithm is "bad" in this sense. They proved that the maximal number of iterations increases exponentially even for a class of linear optimization problems with *nondegenerate* solution sets[33]. Clausen (1987) studied "degenerate modifications" of these solution sets and pointed out that the efficiency of the simplex method is perceptibly worse than in the nondegenerate case. Finally Megiddo (1986) investigated the role of degeneracy in the worst-case complexity of the randomized simplex method. He proved that the problem of "exiting" a degenerate vertex is as difficult as the general linear optimization problem, i.e. if the former is polynomial, so is the latter.

b) Average-case analysis

If the *average* complexity were polynomial, this could explain the "good-natured" behavior of the simplex method in practice. Therefore the average-case was investigated by means of solution sets generated by certain random procedures[34]. Indeed, the average number of simplex iterations turned out to be polynomial, provided that the random procedures are applicable[35]. But a real point of weakness in all these stochastic models is the exclusion (or probability 0) for degeneracy[36]. Further research is necessary to decide whether the average number of iterations grows exponentially if attention is paid to degeneracy[37].

The average case complexity of the randomized simplex method was investigated by Balinski/Liebling/Nobs (1986). They derived an upper bound for the length of a "lexicographic path". The question whether the randomized simplex method is polynomial was not answered definitively.

---

[32] Cf. also Borgwardt(1985:649).

[33] Cf. Klee/Minty (1972:159). Analogous results were found for some variants of the simplex methods (cf. Avis/Chvátal (1978), Goldfarb/Sit (1979), Jeroslow (1973), Zadeh (1973)).

[34] Cf. Borgwardt (1982), Megiddo (1986a), Smale (1983) ,Todd(1986).

[35] Cf. Derigs (1986:55).

[36] Cf. Borgwardt (1987:43).

[37] Cf. Megiddo (1986:367).

2.3 DEGENERACY PROBLEMS WITHIN THE FRAMEWORK OF POSTOPTIMAL ANALYSIS

In sensitivity analysis with respect to the objective function a linear optimization problem

$$\left.\begin{array}{ll} \max z = & c^T x \\ \text{s.t.} & \\ & Ax \quad \leq b \\ & x \quad \geq 0 \end{array}\right\} \qquad (2.3.1)$$

$(x, c \in \mathbb{R}^n, A \in \mathbb{R}^{m\times n}, b \in \mathbb{R}^m)$ and an optimal vertex x^o of (2.3.1) are given[38]. In the simplest case a single component of c is changed, i.e. c is of the form

$$c(t) = c + te_i$$

($e_i \in \mathbb{R}^n, i-$th unit vector; $i \in \{1, \ldots, n\}$ fixed). If x^o is nondegenerate, the problem consists in determining the so-called critical interval $[\underline{t}, \bar{t}]$ with the property that the optimal basis for $t = 0$ remains optimal for each $t \in [\underline{t}, \bar{t}]$. The computation of $\underline{t}$ and $\bar{t}$ is not difficult. The situation is more complicated if x^o is degenerate. In this case several bases correspond to x^o. First of all it is not clear which bases have to remain optimal and which economic facts remain unchanged if the parameter is chosen out of the critical region[39]. The critical interval has to be defined appropriately. Some authors propose the definition[40]

$$T := \cup_{B\in\tilde{B}}[\underline{t}, \bar{t}]^{(B)} \qquad (2.3.2)$$

where $[\underline{t}, \bar{t}]^{(B)}$ denotes the critical interval belonging to matrix B ; $\tilde{B}$ is the set of all primal and dual feasible bases of x^o.

---

[38] Cf. Gal (1987:200ff.).

[39] Cf. Gal (1988:36).

[40] Cf. Piehler (1988), Evans/Baker (1982) and Knolmayer (1984).

With that the "definition-problem" can be regarded as solved, but the economic interpretation of T in (2.3.2) is still a problem[41].

Another problem is the *efficient* determination of T, i.e. the determination of a smallest possible selection of bases $B \in \tilde{B}$, such that the union of the corresponding regions $[\underline{t}, \overline{t}]^{(B)}$ "generates" the whole interval. The only known algorithm that determines T without generating all bases (tableaux) of x^o, was developed by Knolmayer(1984). This procedure generates the bases, performing exclusively dual simplex iterations. Considering the method from the perspective of the theory of optimum graphs[42], certain questions arise.

The theory implies that not any two dual feasible tableaux can be transformed into each other by a sequence of dual simplex iterations. This follows from the fact that there exist disconnected negative optimum graphs. Therefore Knolmayer's algorithm never exits the tableau set belonging to a (maximal connected) component of the negative optimum graph. Thus the explicit proof is missing that the method generates a "complete covering" of T in all cases. An analogous problem arises when Knolmayer's algorithm is used to determine shadow prices under degeneracy[43].

2.4 ON THE PRACTICAL MEANING OF DEGENERACY

Degeneracy occurs very often in practice (cf. Introduction). When a simplex type method enters a degenerate vertex, the determination of the next vertex (or of all neighbouring vertices) is problematic. The possibility of cycling (theoretical infinite repetition of the same sequence of bases, associated with a degenerate vertex) is called a convergence problem, since the optimum is not always reached in a finite number of steps. We speak of an efficiency problem if the termination

---

[41] It is a particularly open question whether the dual infeasible bases, not considered in (2.3.2), should be included from the economical point of view (cf. Piehler (1988, Section 5)).

[42] Cf. Kruse (1987:2, 11).

[43] Cf. Gal (1986).

is insured, but exiting the vertex requires considerable computational effort (e.g. if degeneracy is caused by weak redundancy). In practice it is not possible to distinguish strictly between both kinds of problems: due to rounding errors, "stalling" (cf. Cunningham (1979) and Section 5.1) may occur instead of simplex cycling in the theoretical case, i.e. the simplex method exits the vertex after a finite number of iterations.

Almost all fields of industrial management use mathematical optimization models (cf. Gal (1987), Garfinkel/Nemhauser (1972), Gass (1985), Kiehne (1969) and Varga (1974)):

The *linear model* can be applied to determine the profit-maximizing sales and production program or to solve blending problems (e.g. in the oil- and feed-industry) and cutting stock problems (e.g. in the production of paper and manufacture of sheet iron). Further applications are in agriculture (e.g. determination of a cultivation program with maximal production), financial planning and input-output analysis.

Integer (linear) programming models are used whenever an optimal production program has to be determined with goods appearing in integer quantities only. They further occur in machine load planning, capital budgeting, fixed charge problems and the problem of optimizing utilization of remaining stocks in assembly plants. Finally integer programming appears in the minimization of death duties. *Set covering and partitioning problems* are important special cases of integer programming used especially in airline crew scheduling and route planning. Moreover, *linear and quadratic assignment problems* are special integer programming problems arising in the planning of staff employment and plant and production layout. *Network problems* are used in lowest-cost planning of projects and the determination of maximal flows in networks (e.g. maximal traffic flows in road networks or maximal capacity utilization of gas pipelines).

Vector maximization problems occur in capital budgeting and planning of investment portfolios whenever multiple goals are pursued (e.g. maximization of profit contribution *and* maximization of capacity utilization in the case of production planning). Finally *nonlinear* (especially *quadratic*) *optimization problems* are used in investment planning and pricing.

The above listing shows that degeneracy problems are of considerable importance in practice, since they often involve additional computer time and thus additional costs in the computer-aided solution of planning and decision problems in industrial management.

SUMMARY OF CHAPTER 2

Chapter 2 illustrates that degeneracy may cause diverse problems in almost all kinds of linear constrained mathematical optimization problems whenever a simplex type method is applied. Degeneracy problems arise especially in all fields of optimization in which the linear problem occurs as a subproblem (e.g. in integer programming, parametric programming and methods for finding all vertices of a convex polyhedral set). We have studied mathematical optimization models in which convergence or efficiency problems occur (linear optimization, integer programming, network problems etc.). Degeneracy entails diverse problems with regard to economic interpretation, as well, e.g. the economic meaning of the so-called critical region in sensitivity analysis under degeneracy is not clear. Moreover, the occurence of weakly redundant constraints causes some problems in the interpretation of the optimal solution. Omitting these constraints gives rise to incorrect results in the computation of shadow prices or in sensitivity analysis.

3. THEORY OF DEGENERACY GRAPHS

The present chapter develops additional theory of degeneracy graphs[44], based on the results of Gal (1978,1985) and Kruse (1984,1986).

After the fundamentals have been provided in Section 3.1, we evolve a general theory of $\sigma \times n$-degeneracy graphs ($\sigma, n \in \mathbb{N}$;cf. Section 3.2) which is made possible by set theoretical foundations especially developed for this purpose. In Section 3.3 we study $2 \times n$-degeneracy graphs exclusivly. The "theory of $2 \times n$-degeneracy graphs" is independent of the general theory and is based on the fact that $2 \times n$-degeneracy graphs are special line graphs (cf. Zörnig (1985)).

3.1 FUNDAMENTALS

The following representations are largely based on the publication of Kruse (1986: Ch. 2 and 3), who also provides further details.[45]

3.1.1 THE CONCEPT OF DEGENERACY

Let an $m \times n$-system of linear inequalities of the form[46]

$$\left.\begin{array}{ll} Ax \leq b & (A \in \mathbb{R}^{m \times n}, x \in \mathbb{R}^n, b \in \mathbb{R}^m \\ x \geq 0 & \end{array}\right\} \qquad (3.1.1.1)$$

with the solution set[47]

$$X = \{x \in \mathbb{R}^n | Ax \leq b, x \geq 0\}. \qquad (3.1.1.2)$$

[44] The term "degeneracy graph" was introduced by Kruse (1984,1986), but initial investigations of these graphs were carried out by Gal (1978,1985).

[45] Cf. also Gal (1988:5ff.) and Gal/Kruse/Zörnig (1988).

[46] Cf. Kruse (1986: Section 2, Appendix A).

[47] If we require $b \geq 0$ in (3.1.1.1), X is a convex polyhedral set (cf. Appendix, Def. A.5) with at least one vertex.

be given. Introducing slack variables yields the corresponding canonical form

$$\left.\begin{array}{ll} \bar{A}\bar{x} = b & (\bar{A} = (A|I_m), \bar{x} = \binom{x}{u}) \\ \bar{x} \geq 0 & \\ (I_m \in \mathbb{R}^{m \times m} & \text{(unit matrix)}, \\ u \in \mathbb{R}^m & \text{(vector of slack variables))} \end{array}\right\} \tag{3.1.1.3}$$

with the solution set

$$\bar{X} = \{\bar{x} \in \mathbb{R}^{m+n} | \bar{A}\bar{x} = b, \bar{x} \geq 0\}. \tag{3.1.1.4}$$

Lemma 3.1.1.1[48]**:**

a) The solution sets X and $\bar{X}$ in (3.1.1.2) and (3.1.1.4) are convex polyhedral sets.
b) The mapping $x \mapsto \bar{x} = \binom{x}{u}$ $(u = b - Ax)$ from X to $\bar{X}$ is biunique. The point x is a vertex[49] of X if and only if $\bar{x}$ is a vertex of $\bar{X}$.

A regular $m \times m$-submatrix B of $\bar{A}$ in (3.1.1.3) is a called a basis of $\bar{A}$ (or of (3.1.1.1) or of (3.1.1.3)); $x_B = B^{-1}b \in \mathbb{R}^m$ is called a *basic solution* of (3.1.1.1) for any basis B of $\bar{A}$. If x_B is enlarged by zero components associated with nonbasic variables, the *complete basic solution* $x_B^{(0)} \in \bar{X} \subset \mathbb{R}^{n+m}$ of (3.1.1.3) results. If $x_B^{(0)} \geq 0$, the basis B or the vectors $x_B, x_B^{(0)}$ are called *feasible.*

Lemma 3.1.1.2[50] The complete basic feasible solutions of (3.1.1.3) are the vertices of $\bar{X}$. They are biuniquely assigned to the vertices of X.

Definition 3.1.1.3:

Let a basis B of (3.1.1.1) and the corresponding basic solution $x_B = B^{-1}b$ be given.

a) The basic solution x_B is called *degenerate* if at least one component equals 0.

---

[48] We do not give a proof here, since the statement is fundamental in linear optimization.

[49] Cf. Appendix, Def. A.9.

[50] Cf. Kruse (1986: Appendix, A.14).

b) The number σ of zero components of x_B is called the *degeneracy degree* of x_B $(0 \leq \sigma \leq m)$.
c) A basic solution with degeneracy degree σ is called *σ-degenerate.*

The complete basic solution $x_B^{(0)}$ contains $n + \sigma$ zero components if and only if x_B is σ-degenerate. Therefore the concept of degeneracy is transformable to complete basic (feasible) solutions (vertices of $\bar{X}$) or to the corresponding vertices of X (cf. Lemma 3.1.1.2).

Definition 3.1.1.4:

Let the convex polyhedral sets $X \subset \mathbb{R}^n$ and $\bar{X} \subset \mathbb{R}^{n+m}$ of (3.1.1.2) and (3.1.1.4) be given.

a) A complete basic feasible solution $x_B^{(0)} \in \bar{X}$ is called *degenerate* if the number of its zero components is $n + \sigma$ with $\sigma \geq 1$.
b) A vertex $x^0 \in X$ is called *degenerate* if the corresponding (complete) basic solution is degenerate (cf. Lemma 3.1.1.1b) and Lemma 3.1.1.2).
c) The number σ in a) is called the *degeneracy degree* of $x_B^{(0)}$ or of x^0.
d) A complete basic solution $x_B^{(0)} \in \bar{X}$ or vertex $x_0 \in X$ with degeneracy degree σ is called *σ-degenerate.*

As is well known, the concept of degeneracy can be interpreted geometrically. Let x be any point of X, $\bar{x}$ the corresponding point of $\bar{X}$ (cf. Lemma 3.1.1.1b)). If $\bar{x}$ has a zero component, x lies on a constraint-hyperplane of X (cf. Appendix, Def.A.3). Thus the number of zero components of $\bar{x}$ equals the number of constraint-hyperplanes containing x. Hence a nondegenerate vertex of X is the intersection of exactly n constraint-hyperplanes of X, while a degenerate vertex is contained in more than n of such hyperplanes[51].

To a nondegenerate vertex x^0 or to the corresponding[52] complete basic solution exactly one basis is assigned. In the case of degeneracy

---

[51] Degeneracy can be caused by weak redundancy as well as "proper overdetermination", i.e. no constraint is weak redundant (cf. Kruse (1986:9)).

[52] Cf. Lemma 3.1.1.1b and Lemma 3.1.1.2.

several bases are associated to x^0 (to $x_B^{(0)}$), since b is a linear combination of less than m column vectors of $\bar{A}$ (cf. e.g. Gal (1987:139ff.)).

Definition 3.1.1.5:

Let a (degenerate) complete basic solution $x_B^{(0)} \in \bar{X}$ and the corresponding vertex $x^0 \in X$ be given.

The set B^0 of bases associated with $x_B^{(0)}$ is called the *basis set* of $x_B^{(0)}$ (or of x^0).

3.1.2 THE GRAPHS OF A POLYTOPE

Let the systems and solution sets (3.1.1.1 - 3.1.1.4) be given. Without loss of generality we may assume that X is nonempty and bounded, i.e. X is a convex polytope (cf. Appendix, Def. A. 5). For any two different bases B, B^* let $B \overset{+}{\longleftrightarrow} B^*$ ($B \overset{-}{\longleftrightarrow} B^*; B \leftrightarrow B^*$) denote the possibility of transforming the corresponding tableaux each into the other in one pivot step with a positive (negative, positive or negative) pivot.

We will assign the following graphs to the polytope X (cf. Kruse (1986:11), Grünbaum (1967:212)):

Definition 3.1.2.1:

The graph $G = G(X) = (V, E)$ with

$$V = \{B | B \text{ is a feasible basis of } (3.1.1.1)\}$$

and

$$E = \{\{B, B^*\} \subset V | B \overset{+}{\longleftrightarrow} B^*\}$$

is called the *representation graph*[53] of X.

---

[53] In the definition of E we could substitute $B \overset{-}{\longleftrightarrow} B^*$ or $B \leftrightarrow B^*$ for $B \overset{+}{\longleftrightarrow} B^*$ in order to define further "representation graphs". However, we don't need such modifications in this publication.

Definition 3.1.2.2:
The graph *G'= G'(X) = (V',E')* with

$$V' = \{x \in X | x \text{ is a vertex of X}\}$$

and

$$E' = \{\{x, x^*\} \subset V | x \text{ and } x^* \text{ are connected by an edge of } X\}$$

is called the *graph of the polytope X*.

If all vertices of X are nondegenerate, the graphs *G'(X)* and *G(X)* are isomorphic (cf. Gal 1985:5ff.). The following example illustrates the definitions of *G(X)* and *G'(X)*:

Example 3.1.2.3:

a) Let the system of inequalities (3.1.1.1) be of the special form

$$\left.\begin{array}{rrrl} x_1 & & & \leq 1 \\ & x_2 & & \leq 1 \\ & & x_3 & \leq 1 \\ & & x_1, x_2, x_3 & \geq 0 \end{array}\right\} \qquad (3.1.2.1)$$

The solution set X of (3.1.2.1) has the shape of a cube (cf. Fig. 3.1.2.1.). The graphs *G(X)* and *G'(X)* are isomorphic, since X has no degenerate vertex (cf. Fig. 3.1.2.2). They can be conceived as "projections" of X upon a plane.

b) Now the weak redundant constraint $x_1 + x_2 + x_3 \leq 3$ is added to system (3.1.2.1), i.e we consider the system of inequalities

$$\left.\begin{array}{rrrl} x_1 & & & \leq 1 \\ & x_2 & & \leq 1 \\ & & x_3 & \leq 1 \\ x_1+ & x_2+ & x_3 & \leq 3 \\ & & x_1, x_2, x_3 & \geq 0 \end{array}\right\} \qquad (3.1.2.2)$$

The solution set X' of (3.1.2.2) is identical with the solution set X of (3.1.2.1), but the vertex $x^0 = (1,1,1)^T$ is now degenerate (cf. Fig. 3.1.2.3) Thus the graphs of the polytopes X and X' are isomorphic ($G'(X) \cong G'(X')$), but the representation graph *G(X')* is essentially more complicated than *G(X)* (cf. Fig. 3.1.2.4). The bases (tableaux) associated with the vertex $x^0 = (1,1,1)^T$ are listed in Tab. 3.1.2.1.

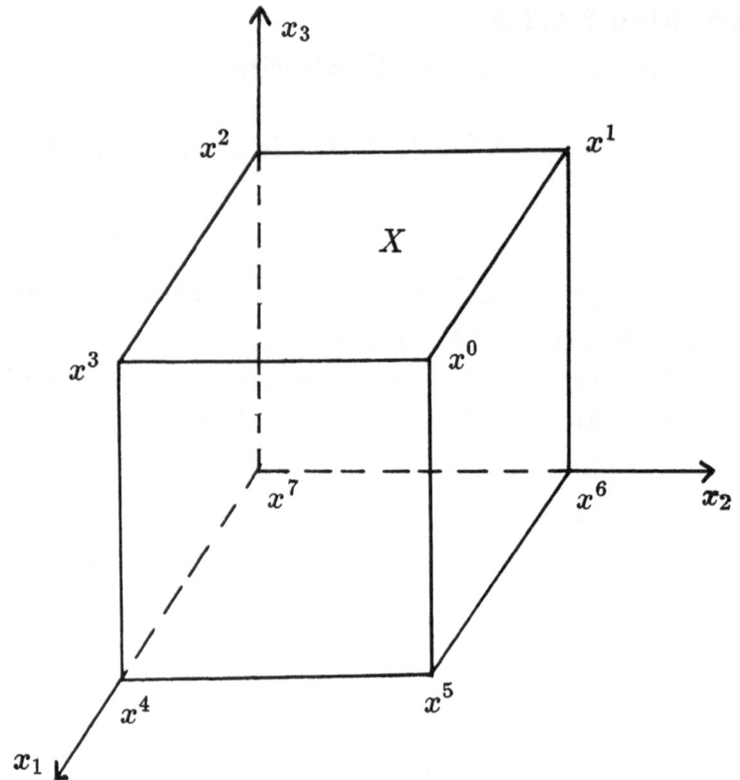

Fig. 3.1.2.1
Solution set X of the system (3.1.2.1)

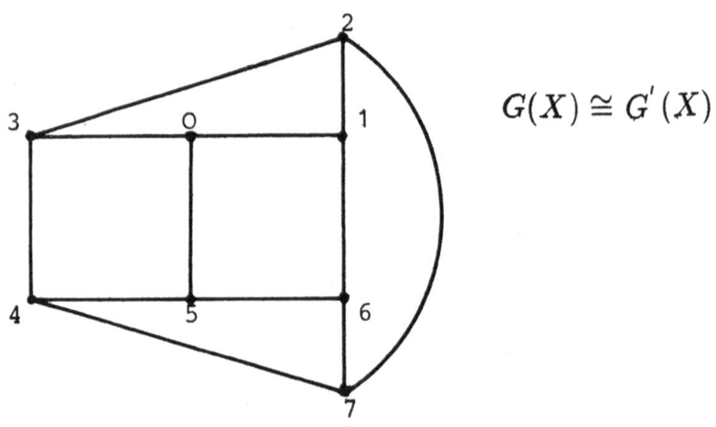

Legend: i - node associated with vertex x^i in Fig. 3.1.2.1 ($i = 0, \ldots, 7$)

Fig. 3.1.2.2
Representation of the graphs $G(X)$, $G'(X)$ associated with the polytope X in Fig. 3.1.2.1

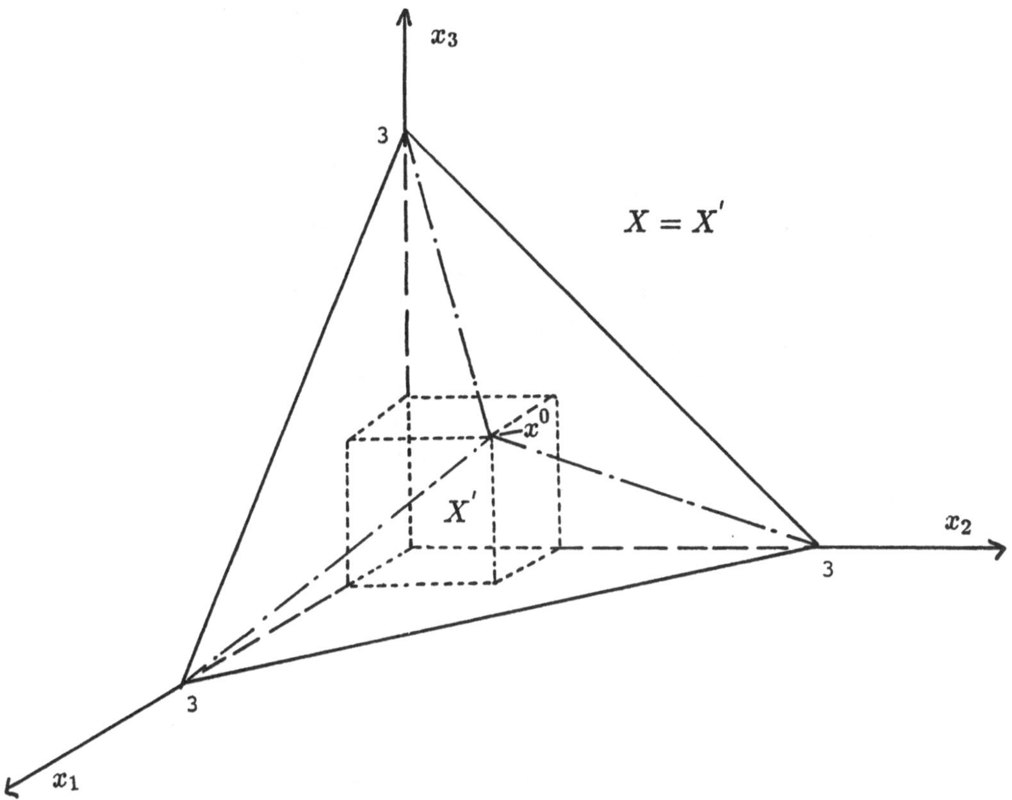

Fig 3.1.2.3
Solution set X' of system (3.1.2.2)

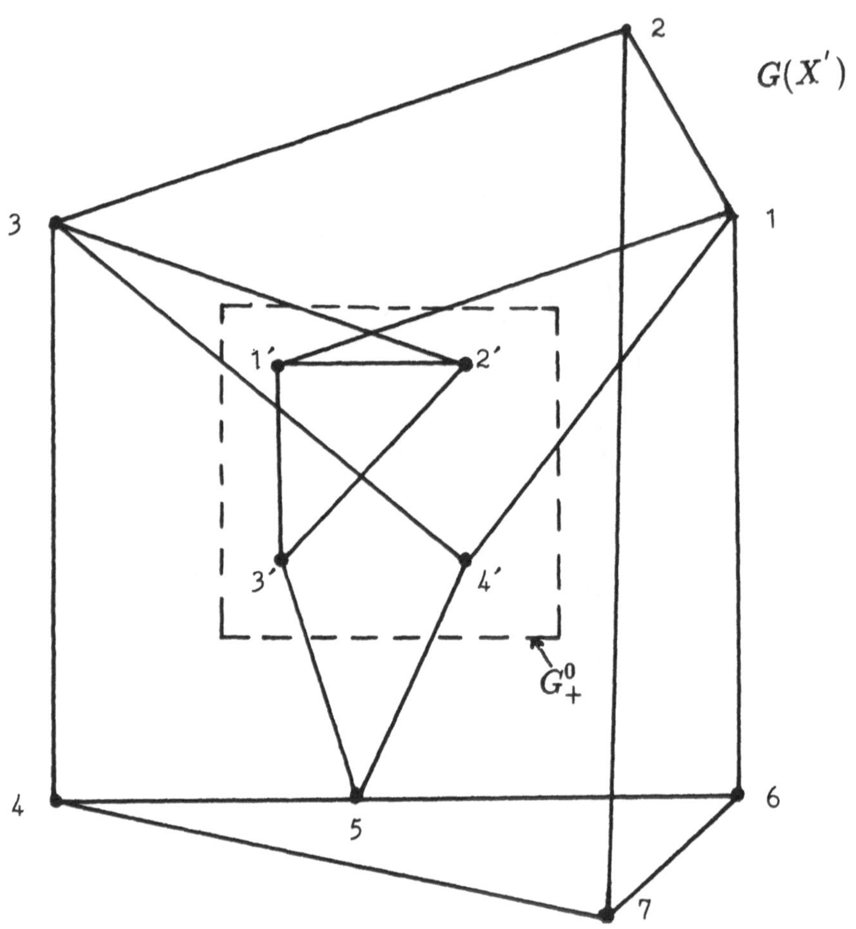

Legend:

i - node (basis) of the nondegenerate vertex x^i (i = 1,. . .,7)

u'- node (basis) B^0_u of the basis set $B^0=\{B^0_1,...,B^0_4\}$
of the degenerate vertex $x^0=(1,1,1)^T \in X'$ (cf. Tab. 3.1.2.1)

Fig. 3.1.2.4
Representation graph $G(X')$ of the polytope X' in Fig. 3.1.2.3

Tab. 3.1.2.1
Bases and reduced tableaux of the degenerate vertex $\mathbf{x^0 = (1, 1, 1)^T}$ of $\mathbf{X'}$ in Fig. 3.1.2.3.

B_1^0	5	6	7	
1	-1	-1	1	1
2	1	0	0	1
3	0	1	0	1
4	1	1	-1	0

B_2^0	4	6	7	
1	1	0	0	1
2	-1	-1	1	1
3	0	1	0	1
5	1	1	-1	0

B_3^0	4	5	7	
1	1	0	0	1
2	0	1	0	1
3	-1	-1	1	1
6	1	1	-1	0

B_4^0	4	5	6	
1	1	0	0	1
2	0	1	0	1
3	0	0	1	1
7	-1	-1	-1	0

Legend: Tableau associated with basis B_u^0 $(u = 1, \ldots, 4)$

B_u^0	column indices	
Basic indices	nonbasic columns	basic solution

The example above demonstrates that for degenerate vertices "information will be lost" to some extent, if a polytope is presented by the graph of a polytope (cf. Def. 3.1.2.2). On the other hand the representation graph reflects the structure of *all* vertices, even the degenerate ones.

3.1.3 DEGENERACY GRAPHS

Let (3.1.1.1) - (3.1.1.4) be given again. Moreover, let $x^0 \in X$ be a σ-degenerate vertex with the corresponding basis set B^0.

The following graphs represent the structure of x^0 (cf. Kruse (1986)):

Definition 3.1.3.1:

The graph $G^0 = (B^0, E^0)$ with the edge set

$$E^0 = \{\{B, B^*\} \subset B^0 | B \leftrightarrow B^*\}$$

is called the *(general) degeneracy graph* of x^0.

The graph $G^0_+ = (B^0, E^0_+)$ with the edge set

$$E^0_+ = \{\{B, B^*\} \subset B^0 | B \overset{+}{\longleftrightarrow} B^*\}$$

is called the *positive degeneracy* graph of x^0 and the graph $G^0_- = (B^0, E^0_-)$ with the edge set

$$E^0_- = \{\{B, B^*\} \subset B^0 | B \overset{-}{\longleftrightarrow} B^*\}$$

is called the *negative degeneracy* graph of x^0.

Degeneracy graphs with the given degeneracy degree σ and the given dimension n are called $\sigma \times n$*-degeneracy graphs.*

The graphs G^0_- and G^0_+ are complementary subgraphs[54] of G^0 (cf. Fig. 3.1.3.1)

The following (simple) example illustrates degeneracy graphs and their interrelations[55].

Example 3.1.3.2:

Consider the degenerate vertex $x^0 = (1, 1, 1)^T$ of the solution set X' of (3.1.2.2). The corresponding basis set is $B^0 = \{B_1, \ldots, B_4\}$ (cf. Tab. 3.1.2.1). Fig. 3.1.3.1 represents the degeneracy graphs of x^0.

Comparing Fig. 3.1.3.1 and 3.1.2.4 illustrates that the positive degeneracy graph of $x^0 \in X$ is an induced subgraph[56] of the representation graph *G(X)*.

The following statements are necessary for a detailed theoretical investigation of degeneracy graphs (cf. Kruse (1986)).

After appropriately exchanging rows and columns and "re-indexing" a pivot tableau assigned to a degenerate vertex can be represented in the form of Tab. 3.1.3.1 or in the corresponding reduced form (Tab. 3.1.3.2).

---

[54] Cf. Appendix, Def. B.3.

[55] Kruse (1986:13) presents a more complicated positive degeneracy graph (note the area with hatching).

[56] Cf. Appendix, Def. B.3.

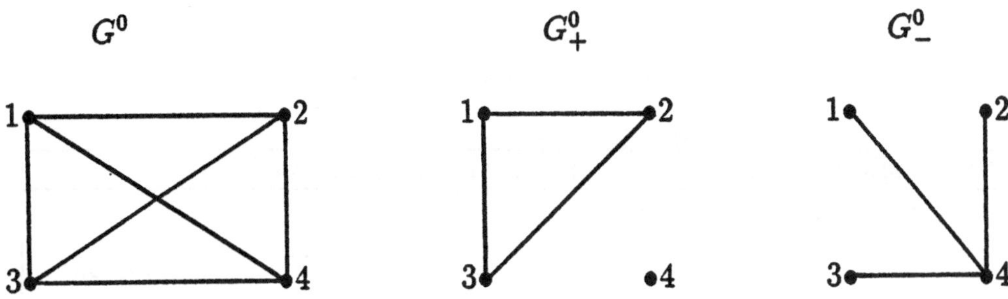

Legend: u – node (basis) B^0_u (u=1,...,4).

Fig. 3.1.3.1
The degeneracy graphs G^0, G^0_+, G^0_- of the degenerate vertex $\mathbf{x}^0 = (1,\ 1,\ 1)^T$ of the solution set X' of (3.1.2.2).

Tab. 3.1.3.1
Pivot tableau of a basis of the σ-degenerate vertex x^0.

column indices

basic-indices	$1 \dots \sigma$	$\sigma+1 \dots$ m	$m+1$	$\dots$ $m+n$	
1	1	0 … 0	$y_{1,m+1}$	$y_{1,m+n}$	0
⋮	⋱	⋮ ⋮	⋮	⋮	⋮
σ	1	0 … 0	$y_{\sigma,m+1}$	$y_{\sigma,m+n}$	0
$\sigma+1$	0 0	1	$y_{\sigma+1,m+1}$	$y_{\sigma+1,m+n}$	$y_{\sigma+1}$
⋮	⋮ ⋮	⋱	⋮	⋮	⋮
m	0 … 0	1	$y_{m,m+1}$	$y_{m,m+n}$	y_m

Definition 3.1.3.3:

Let a pivot tableau of a σ-degenerate vertex x^0 be given in the form of Tab. 3.1.3.2.

a) The subtableau in bold type is called a *degeneracy tableau* of x^0.

b) The following system of linear inequalities, defined by a degeneracy tableau, is called a canonical system of x^0.

$$\left.\begin{array}{l} Yx \leq 0 \quad (Y \in \mathbb{R}^{\sigma \times n}, x \in \mathbb{R}^n) \\ x \geq 0 \end{array}\right\} \qquad (3.1.3.1)$$

Tab. 3.1.3.2
Reduced form of Tab 3.1.3.1

	column indices			
basic indices	$1\dots\sigma$	$\sigma+1\dots m$	$m+1\dots m+n$	
1 $\vdots$ σ	I_σ	$O_{\sigma\times(m-\sigma)}$	$Y = Y_{\sigma\times n}$	$O\in \mathbb{R}^\sigma$
$\sigma+1$ $\vdots$ m	$O_{(m-\sigma)\times\sigma}$	$I_{m-\sigma}$	$Y_{(m-\sigma)\times n}$	$y \in \mathbb{R}^{m-\sigma}$

Obviously $\tilde{x}^0 = 0 \in \mathbb{R}^n$ is a σ-degenerate vertex of $\tilde{X}$, where $\tilde{X}$ denotes the solution set of (3.1.3.1)[57].

Definition 3.1.3.4:

Let the canonical system (3.1.3.1) of a degenerate vertex x^0 with solution set $\tilde{X}$ be given. The general (positive, negative) degeneracy graph of the vertex $\tilde{x}^0 = 0 \in \tilde{X}$ is called the *general (positive, negative) canonical degeneracy graph* of x^0 and is denoted by $G_Y(G_Y^+, G_Y^-)$.

Remark 3.1.3.5:[58] The degeneracy graphs G^0, G_+^0, G_-^0 of a degenerate vertex x^0 are isomorphic to the corresponding canonical degeneracy graphs:

$$G^0 \cong G_Y, G_+^0 \cong G_Y^+, G_-^0 \cong G_Y^-.$$

So far as isomorphic graphs are identified with each other, every degeneracy graph is of the form G_Y, G_Y^+ or G_Y^-.

The following assertion is useful for the investigation of *general*[59] degeneracy graphs.

---

[57] $\tilde{x}^0$ is the unique vertex of $\tilde{X}$, since $\tilde{X}$ is a convex cone (cf. Appendix, Lemma A.8).

[58] Cf. Kruse (1986:30).

[59] Cf. Def. 3.1.3.1.

Remark 3.1.3.6:

The nodes of G_Y correspond to bases (regular $\sigma \times \sigma$-submatrices) of $(Y|I_\sigma)$. Two nodes (bases) of G_Y are neighbouring if and only if they differ in exactly one column.

In order to simplify formulations we frequently identify a node (basis) of G_Y with the corresponding index set I.[60]

In connection with explaining simplex cycling we will introduce further degeneracy graphs, namely LP-degeneracy graphs (cf. Section 4.2.1). Moreover, investigations in the following fields have led to the definition of other kinds of degeneracy graphs or related typs of graphs (cf. Altherr (1975), Gal (1979:23ff.), Geue (1989: Section 4), Kruse (1986:10, 1987:3), Jansson (1985:15ff., 83ff.), Piehler/Kruse (1989)):

- sensitivity analysis/ determination of shadow prices under degeneracy
- development of a new pivoting rule
- determination of all vertices of a polytope

[60] This simplifies formulations e.g. in the proof of Theorem 3.2.3.6. Occasionally the term "basic index" is used for the index set (cf. e.g. Gal (1985:581)).

3.2 THEORY OF $\sigma \times n$-DEGENERACY GRAPHS

First of all we develop some concepts and interrelations on the theory of finite sets, which are fundamental for the characterization of $\sigma \times n$-degeneracy graphs (cf. Section 3.2.2).

3.2.1 FOUNDATIONS OF THE THEORY OF FINITE SETS

In this section σ, N denote any natural numbers with $\sigma < N$.

Definition 3.2.1.1:
Let $\mathcal{S} := \{s_1, \ldots s_p\}$ be a set system on $\{1, \ldots, N\}$. System $\mathcal{S}$ is called *σ-homogeneous* if $|s_j| = \sigma$ holds for $j = 1, \ldots, p$.

Example 3.2.1.2:
The system

$$\mathcal{S} := \{\{1,2,3\}, \{1,5,6\}, \{2,3,4\}, \{2,5,6\}, \{3,4,5\}\}$$

is a 3-homogeneous system on $\{1, \ldots, 6\}$. It is illustrated in Fig. 3.2.1.1.

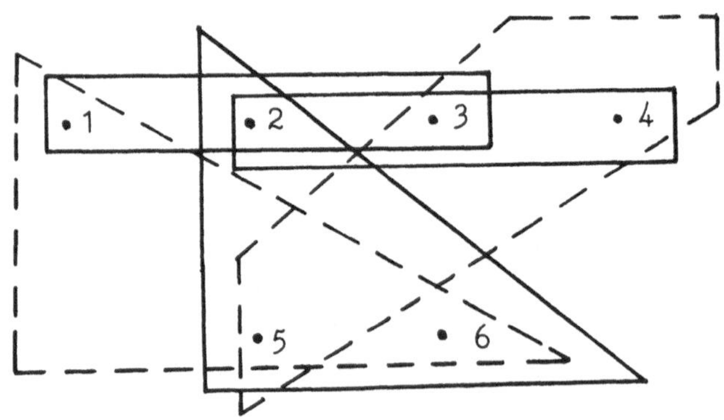

Fig. 3.2.1.1
Illustration of the 3-homogeneous set system $\mathcal{S}$ on $\{1, \ldots, 6\}$ in Example 3.2.1.2

Definition 3.2.1.3:

Let $\mathcal{S} := \{s_1, \dots, s_p\}$ be a σ-homogeneous system on $\{1, \dots, N\}$. The *complementary (σ-homogeneous) system* $\bar{\mathcal{S}}$ of $\mathcal{S}$ is defined by

$$\bar{\mathcal{S}} := \{s \subset \{1, \dots, N\} \mid \ |s| = \sigma, s \notin \mathcal{S}\}.$$

Obviously the complementary system of $\bar{\mathcal{S}}$ is $\mathcal{S}$, i.e $\mathcal{S} = \bar{\bar{\mathcal{S}}}$.

Example 3.2.1.4:

The complementary system of $\mathcal{S}$ in Example 3.2.1.2 is

$$\begin{aligned}\bar{\mathcal{S}} = \{&\{1,2,4\}, \{1,2,5\}, \{1,2,6\}, \{1,3,4\}, \{1,3,5\},\\ &\{1,3,6\}, \{1,4,5\}, \{1,4,6\}, \{2,3,5\}, \{2,3,6\},\\ &\{2,4,5\}, \{2,4,6\}, \{3,4,6\}, \{3,5,6\}, \{4,5,6\}\}\end{aligned}$$

Instead of listing all sets of a σ-homogeneous system $\mathcal{S}$, we can "represent" $\mathcal{S}$ by formulating a system $\mathcal{T} = \{t_1, \dots, t_q\}$ on $\{1, \dots, N\}$ ($|t_i| \geq \sigma, i = 1, \dots, q$), such that all σ-subsets of the sets $t_1, \dots, t_q$ belong to $\mathcal{S}$. This idea motivates

Definition 3.2.1.5:

Let $\mathcal{S} := \{s_1, \dots, s_p\}$ be a σ-homogeneous system on $\{1, \dots, N\}$. A system $\mathcal{T} := \{t_1, \dots, t_q\}$ on $\{1, \dots, N\}$ with $|t_i| \geq \sigma$ for $i = 1, \dots, q$ is called a *representation system* of $\mathcal{S}$ if

$$\mathcal{S} = \{s \subset \{1, \dots, N\} \mid \ |s| = \sigma, s \subset t_i \text{ for at least one } i \in \{1, \dots, q\}\}.$$

Vice versa, $\mathcal{S}$ is called the *σ-homogeneous system, represented by $\mathcal{T}$*.

Hence a system $\mathcal{T} = \{t_1, \dots, t_q\}$ ($|t_i| \geq \sigma$ for $i = 1, \dots, q$) is a representation system of $\mathcal{S}$, when $\mathcal{S}$ consists of all σ-subsets of the sets $t_1, \dots, t_q$. Vice versa, system $\mathcal{T}$ represents the system of all σ-subsets contained in any set $t_1, \dots, t_q$.

We illustrate the definition by

Example 3.2.1.6:

It is easily proved that

$$\begin{aligned}\mathcal{T} =&\{\{1,2,4,5\},\{1,2,4,6\},\{1,3,4,6\},\{1,3,5\},\{2,3,5\},\\&\{2,3,6\},\{3,5,6\},\{4,5,6\}\}\end{aligned}$$

and

$$\begin{aligned}\mathcal{T}' =&\{\{1,2,4,5\},\{1,2,4,6\},\{1,3,4\},\{1,3,6\},\{1,4,6\},\\&\{3,4,6\},\{1,3,5\},\{2,3,5\},\{2,3,6\},\{3,5,6\},\{4,5,6\}\}\end{aligned}$$

are both representation systems of the set system $\bar{\mathcal{S}}$ in Example 3.2.1.4.

The example implies that different representation systems may exist for a given σ-homogeneous set system. However, for the following considerations it is important that σ-homogeneous systems can be biuniquely assigned to representation systems. For this purpose it is necessary to introduce a "normal form" of a representation system. This is done in

Definition 3.2.1.7:

Let $\mathcal{S} := \{s_1, \ldots, s_p\}$ be a σ-homogeneous system on $\{1, \ldots, N\}$. The system $\tilde{\mathcal{D}}(\mathcal{S})$ is defined by

$$\tilde{\mathcal{D}}(\mathcal{S}) := \{t \subset \{1, \ldots, N\} | \quad |t| \geq \sigma; s \subset t, |s| = \sigma \Rightarrow s \in \mathcal{S}\}.$$

The system of maximal sets of $\tilde{\mathcal{D}}(\mathcal{S})$ (with respect to inclusion) is denoted by $\mathcal{D}(\mathcal{S})$; $\mathcal{D}(\mathcal{S})$ is called the *σ-normal representation system of $\mathcal{S}$*.

Example 3.2.1.8:

Consider the 3-homogeneous system $\bar{\mathcal{S}}$ of Example 3.2.1.4. According to Definition 3.2.1.7 holds

$$\tilde{\mathcal{D}}(\bar{\mathcal{S}}) := \{\{1,2,4,5\},\{1,2,4,6\},\{1,3,4,6\}\} \cup \bar{\mathcal{S}}$$

System $\mathcal{D}(\bar{\mathcal{S}})$ results from this by omitting all 3-subsets of $\tilde{\mathcal{D}}(\bar{\mathcal{S}})$ contained in any of the 4-subsets of $\tilde{\mathcal{D}}(\bar{\mathcal{S}})$, i.e.

$$\begin{aligned}\mathcal{D}(\bar{\mathcal{S}}) =&\{\{1,2,4,5\},\{1,2,4,6\},\{1,3,4,6\},\{1,3,5\},\{2,3,5\},\\ &\{2,3,6\},\{3,5,6\},\{4,5,6\}\}.\end{aligned}$$

Hence, the σ-normal representation system of $\bar{\mathcal{S}}$ coincides with system $\mathcal{T}$ in Example 3.2.1.6.

Definition 3.2.1.9:

A set system $\mathcal{T}$ on $\{1,\ldots,N\}$ is called $\sigma - \textit{normal}$ if $\mathcal{T} = \mathcal{D}(\mathcal{S})$ holds for any σ-homogeneous system $\mathcal{S}$ on $\{1,\ldots,N\}$.

Thus a set system $\mathcal{T}$ on $\{1,\ldots,N\}$ is σ-normal if it is the σ-normal representation system of any σ-homogeneous system $\mathcal{S}$ on $\{1,\ldots,N\}$. For example the above system $\mathcal{T}$ is a 3-normal system (cf. Example 3.2.1.6).

The interrelations between the above concepts are established in

Lemma 3.2.1.10:

The mapping

$$\mathcal{D} : \mathcal{S} \mapsto \mathcal{T} := \mathcal{D}(\mathcal{S})$$

from σ-homogeneous set systems to σ-normal set systems is biunique.

Proof: Let $\mathcal{S} := \{s_1,\ldots,s_p\}$, $\mathcal{S}' := \{s_1',\ldots,s_q'\}$ be two *distinct* σ-homogeneous set systems on $\{1,\ldots,N\}$. Without loss of generality we may assume $s_1 \notin \mathcal{S}'$.

So $\mathcal{D}(\mathcal{S})$ contains a set, a subset of which is s_1. Since this does not hold for $\mathcal{D}(\mathcal{S}')$, it follows that $\mathcal{D}(\mathcal{S}) \neq \mathcal{D}(\mathcal{S}')$.•

Definition 3.2.1.11:

Let $Y \in \mathbb{R}^{\sigma \times n}$ be a matrix which is feasibly laid, i.e Y has neither zero rows nor zero columns (cf. Kruse (1986:40)). Moreover, let $t_1, \ldots, t_q$ denote the index sets of maximal $\sigma \times k$- submatrices of $(Y|I_\sigma)$ with rank $< \sigma$ ($I_\sigma \in \mathbb{R}^{\sigma \times \sigma}$ unit matrix, $\sigma \leq k \leq n + \sigma$).

The set system $\mathcal{T} := \{t_1, \ldots, t_q\}$ on $\{1, \ldots, N\}$ ($N := n + \sigma$) is called the *system induced by* Y and is denoted by $\mathcal{T}_Y$.

Example 3.2.1.12:

Consider the matrix

$$Y = \begin{pmatrix} 2 & 2 & 3 & 0 & 0 \\ 2 & 0 & 0 & 2 & 3 \end{pmatrix} \in \mathbb{R}^{2 \times 5} \quad (\sigma = 2, n = 5, N = 7).$$

The column vectors $y^1, \ldots, y^7$ of $(Y|I_2)$ are presented graphically in Fig. 3.2.1.2.

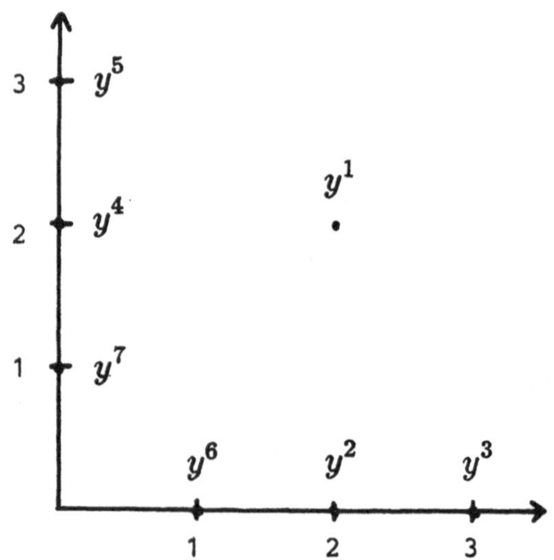

Fig. 3.2.1.2
Column vectors y^j of $(Y \mid I_2)$ ($j = 1, \ldots, 7$)

Let us consider e.g. the 2×3-submatrix $M = (y^4|y^5|y^7)$ of $(Y|I_2)$. Fig. 3.2.1.2 shows that the column vectors of M are points of a straight line passing the origin. Thus rank $M = 1$, i.e. $\{4,5,7\}$ is an index set of a 2×3-submatrix of $(Y|I_2)$ with rank < 2 (namely of M). Analogous considerations for all possible choices of column vectors result in the following collection of $2 \times k$-submatrices of $(Y|I_2)$ with rank < 2 $(k \leq 7)$:

$$\{2,3,6\},\{2,3\},\{2,6\},\{3,6\},\{4,5,7\},\{4,5\},\{4,7\},\{5,7\}.$$

Hence the system on $\{1,\ldots,7\}$ induced by Y is

$$\mathcal{T} = \{\{2,3,6\},\{4,5,7\}\}.$$

Definition 3.2.1.13:

A set system $\mathcal{T}$ on $\{1,\ldots,N\}$ is called *σ-induced* if $\mathcal{T} = \mathcal{T}_Y$ holds for any feasibly laid matrix $Y \in \mathbb{R}^{\sigma \times n}$ $(N = n + \sigma)$.

Finally we point out some properties of σ-homogeneous and σ-induced set systems.

Theorem 3.2.1.14:[61]

Every σ-induced set system $\mathcal{T}$ on $\{1,\ldots,N\}$ is σ-normal $(N = n + \sigma)$.

Proof: Let a σ-induced set system $\mathcal{T}$ on $\{1,\ldots,N\}$ and a matrix $Y \in \mathbb{R}^{\sigma \times n}$ be given, such that $\mathcal{T} = \mathcal{T}_Y$ $(N = n + \sigma)$. Let $\mathcal{S}$ be the σ-homogeneous set system represented by $\mathcal{T}$, i.e. let $\mathcal{S}$ be the system of index sets of singular $\sigma \times \sigma$-submatrices of $(Y|I_\sigma)$.

From Definition 3.2.1.7 follows $\mathcal{D}(\mathcal{S}) = \mathcal{T} = \mathcal{T}_Y$, i.e. $\mathcal{T}$ is σ-normal.•

With regard to the theory of $\sigma \times n$-degeneracy graphs it would be desirable to have a "simple" necessary and sufficient criterion for the "σ-inducedness" of set systems (cf. the considerations following Example 3.2.2.9).

The following theorem represents a first step in this direction. It provides necessary conditions for "σ-inducedness" in the case $\sigma = 3$.

[61] Cf. Lemma 3.2.1.16.

Theorem 3.2.1.15:

Let $\mathcal{T} = \mathcal{T}_Y = \{t_1, \ldots, t_q\}$ be a 3-induced set system on $\{1, \ldots, N\}$ ($Y \in \mathbb{R}^{\sigma \times n}, N = n + \sigma; \sigma = 3; Y$ feasibly laid). Then $\mathcal{T}$ satisfies the following condition :

For all $Q' \subset Q \subset \{1, \ldots, q\}, |Q'| \geq 2$ holds:

$$\cap_{i \in Q} t_i \neq \emptyset \Rightarrow \cap_{i \in Q'} t_i = \cap_{i \in Q} t_i$$

Proof: Let Q, Q' be sets with $Q' \subset Q \subset \{1, \ldots, q\}$, $|Q'| \geq 2$ and let be

$$\cap_{i \in Q} t_i \neq \emptyset. \tag{3.2.1.1}$$

We have to show that

$$\cap_{i \in Q'} t_i = \cap_{i \in Q} t_i \tag{3.2.1.2}$$

holds.

Let U_i denote the subspace of $\mathbb{R}^3$, generated by the column vectors of $(Y|I_3)$, associated with t_i $(i = 1, \ldots, q)$.

Since the index sets t_i are associated with maximal matrices of rank < 3, the subspaces U_i represent pairwise disjoint planes in the three-dimensional space, containing the origin (cf. Def. 3.2.1.11).

From (3.2.1.1) follows

$$k \in \cap_{i \in Q} t_i$$

for any index $k \in \{1, \ldots, N\}$. Thus

$$0, y^k \in \cap_{i \in Q} U_i, \tag{3.2.1.3}$$

where $y^k \neq 0$ denotes the k-th column vector of $(Y|I_3)$. Because of (3.2.1.3) the planes U_i contain the straight line $G_k := \text{span}\{y^k\}$, hence

$$\cap_{i \in Q} U_i = \cap_{i \in Q'} U_i. \tag{3.2.1.4}$$

Since $t_i = \{j \in \{1,\ldots,N\} | y^j \in U_i\}$, (3.2.1.4) implies

$$\begin{aligned}\cap_{i\in Q} t_i &= \{j \in \{1,\ldots,N\} | y^j \in \cap_{i\in Q} U_i\} \\ &= \{j \in \{1,\ldots,N\} | y^j \in \cap_{i\in Q'} U_i\}, \\ &= \cap_{i\in Q'} t_i\end{aligned}$$

i.e (3.2.1.2) holds and the proof is completed.•

From the result above follows that the reverse assertion of Theorem 3.2.1.14 does not hold:

Lemma 3.2.1.16

A σ-normal set system on $\{1,\ldots,N\}$ is not necessarily σ-induced $(N = n + \sigma)$.

Proof: Consider the system $\mathcal{T} = \{t_1, t_2, t_3\}$ on $\{1,\ldots,6\}$ with $t_1 = \{1,2,4\}$, $t_2 = \{2,3,4,5\}$, $t_3 = \{4,5,6\}$.
$\mathcal{T}$ is 3-normal but not 3-induced $(\sigma = n = 3, N = 6)$.

a) $\mathcal{T} = \mathcal{D}(\mathcal{S})$ mit $\mathcal{S} = \{\{1,2,4\},\{2,3,4\},\{2,3,5\},\{2,4,5\},\{3,4,5\},\{4,5,6\}\}$ (cf. Def. 3.2.1.7), i.e. $\mathcal{T}$ is 3-normal.
b) For $Q' = \{1,2\}, Q = \{1,2,3\}$ holds $\cap_{i\in Q} t_i = \{4\} \neq \emptyset$ and $\cap_{i\in Q'} t_i = \{2,4\} \neq \cap_{i\in Q} t_i = \{4\}$.

Hence the condition in Theorem 3.2.1.15 is not satisfied for $\mathcal{T}$, i.e. $\mathcal{T}$ is not 3-induced.•

3.2.2 CHARACTERIZATION OF $\sigma \times n$-DEGENERACY GRAPHS

The concepts of Section 3.2.1 make a characterization of $\sigma \times n$-degeneracy graphs possible $(\sigma \geq 2)$. This is an assertion of the following form:

"A graph is a $\sigma \times n$-degeneracy graph if and only if it has the properties..."

Such a characterization yields a deeper insight into the structure of degeneracy graphs[62]. In particular, it can be used to decide whether a given graph is a degeneracy graph[63]. Based on the characterization we will derive certain structural properties of degeneracy graphs (e.g. numbers of nodes and vertices, connectivity).[64]

In order to characterize degeneracy graphs, we have to embed them at first into the more general class of index graphs (cf. Def. 3.2.2.2). After that we will assign biuniquely a certain σ-homogeneous system, namely the so-called representation system, to every $\sigma \times n$-index graph (cf. Def. 3.2.2.6).

Definition 3.2.2.1:

For $\sigma, n \in I\!N, N := n + \sigma$ we define the graph $G^{\sigma,n}$ as follows:

a) The nodes of $G^{\sigma,n}$ correspond to σ-subsets of $\{1,\ldots,N\}$.
b) Two nodes of $G^{\sigma,n}$ are connected by an edge if and only if the σ-subsets have $(\sigma - 1)$ elements in common.

The graph $G^{\sigma,n}$ is called the *complete $\sigma \times n$-index graph.*

Definition 3.2.2.2:

The induced subgraphs[65] $\langle \mathcal{S} \rangle$ of $G^{\sigma,n}$ are called *$\sigma \times n$-index graphs.*

Example 3.2.2.3:

The complete 2×3-index graph $G^{2,3}$ and the 2×3-index graph, induced by the node set $\mathcal{S} = \{\{1,4\},\{2,4\},\{2,5\},\{3,4\},\{4,5\}\}$ are illustrated in Fig. 3.2.2.1.

Remark 3.2.2.4:

We conceive two index graphs $\langle \mathcal{S}_1 \rangle$, $\langle \mathcal{S}_2 \rangle$ as different if $\mathcal{S}_1 \neq \mathcal{S}_2$ holds for the node sets. The example above shows that different index graphs

---

62 Characterizations are already known for other types of graphs, e.g. for line graphs (cf. Beineke/Wilson (1978:278ff.)), for Eulerian graphs and planar graphs (cf. Behzad/Chartrand (1971:35ff, 90ff.)).

63 Cf. Examples 3.2.2.9 and 3.2.2.11.

64 Cf. Sections 3.2.3 and 3.3.2.

65 Cf. Appendix, Def. B.3.

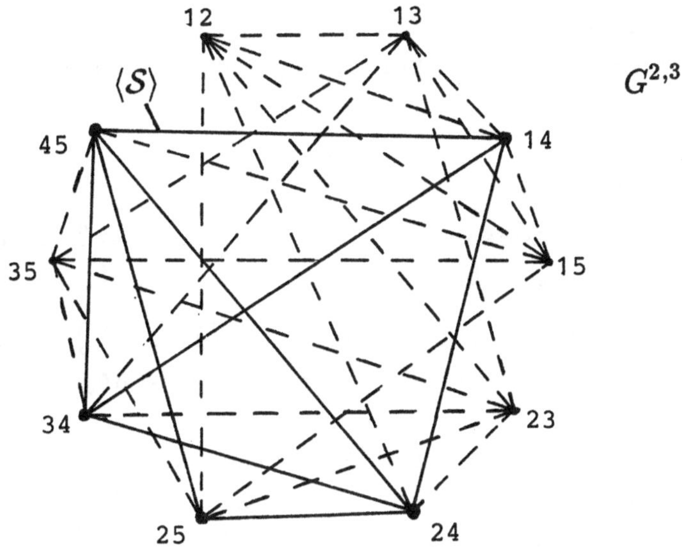

Legend:
i, j – node $\{i,j\}$ ($\{i,j\} \subset \{1,\ldots,5\}$)
- – – – – edge of $G^{2,3}$
—— – edge of $\langle S \rangle$

Fig. 3.2.2.1
Representation of $G^{2,3}$ and $\langle S \rangle$ in Example 3.2.2.3

may be isomorphic, e.g. the subgraph of $G^{2,3}$, induced by the node set $\{\{1,2\},\{2,4\},\{4,5\},\{2,3\},\{2,5\}\}$ is isomorphic to $\langle S \rangle$ (cf. Fig. 3.2.2.1).

Remark 3.2.2.5:

We identify the $\sigma \times n$-degeneracy graphs with the $\sigma \times n$-index graphs induced by the index sets of bases (regular $\sigma \times \sigma$-submatrices) of a matrix $(Y|I_\sigma)$ ($Y \in \mathbb{R}^{\sigma \times n}$; cf. Remarks 3.1.3.5 and 3.1.3.6).[66]

Let $\mathcal{S}$ be a subset of the node set of $G^{\sigma,n}$. The set $\mathcal{S}$ can be biuniquely assigned to the subgraph $\langle \mathcal{S} \rangle$ of $G^{\sigma,n}$, induced by $\mathcal{S}$. Moreover,

[66] Note that this identification enables us to distinguish between isomorphic degeneracy graphs or index graphs (cf. Remark 3.2.2.4).

$\mathcal{S}$ is a σ-homogeneous system on $\{1,\ldots,N\}$ $(N=n+\sigma)$ which can be biuniquely assigned to a σ-normal representation system $\mathcal{T}=\mathcal{D}(\bar{\mathcal{S}})$.[67]

This idea motivates the following

Definition 3.2.2.6:

Let $\langle\mathcal{S}\rangle$ be a $\sigma\times n$-index graph. The set system $\mathcal{T}=\mathcal{D}(\bar{\mathcal{S}})$ on $\{1,\ldots,N\}$ $(N=n+\sigma)$ is called the *representation system* of $\langle\mathcal{S}\rangle$.

Thus a representation system $\mathcal{T}$ on $\{1,\ldots,N\}$ is biuniquely assigned to any $\sigma\times n$-index graph, especially to any $\sigma\times n$-degeneracy graph $(N=n+\sigma)$.

So $\mathcal{T}$ contains "all information" about $\langle\mathcal{S}\rangle$ in a certain sense.

In spite of this, $\mathcal{T}$ may be essentially more "simple" in structure than $\langle\mathcal{S}\rangle$, as the following examples demonstrate.

Example 3.2.2.7:

Let be $\sigma=n=3, N=6$.

a) We consider the 3×3-index graph $\langle\mathcal{S}\rangle$ with node set $\mathcal{S}=\{\{1,3,5\},\{1,3,6\},\{1,4,5\},\{1,4,6\},\{2,3,5\},\{2,3,6\},\{2,4,5\},\{2,4,6\}\}$. The graph $\langle\mathcal{S}\rangle$ is illustrated in Fig. 3.2.2.2. The complementary system is: $\bar{\mathcal{S}}=\{\{1,2,3\},\{1,2,4\},\{1,2,5\},\{1,2,6\},\{1,3,4\},\{1,5,6\},\{2,3,4\},\{2,5,6\},\{3,4,5\},\{3,4,6\},\{3,5,6\},\{4,5,6\}\}$.
Thus the representation system of $\langle\mathcal{S}\rangle$ is $\mathcal{D}(\bar{\mathcal{S}})=\{\{1,2,3,4\},\{3,4,5,6\},\{1,2,5,6\}\}$.

b) We consider the 3×3-index graph $\langle\mathcal{S}\rangle$ with $\mathcal{S}=\{\{1,2,3\},\{1,2,5\},\{1,2,6\},\{1,3,4\},\{1,3,5\},\{1,3,6\},\{1,4,5\},\{1,4,6\},\{1,5,6\},\{2,3,6\},\{2,4,6\},\{2,5,6\},\{3,4,6\},\{3,5,6\}\}$. The graph $\langle\mathcal{S}\rangle$ is illustrated in Fig. 3.2.2.3. The complementary system is $\bar{\mathcal{S}}=\{\{1,2,4\},\{2,3,4\},\{2,3,5\},\{2,4,5\},\{3,4,5\},\{4,5,6\}\}$.
Thus the representation system of $\langle\mathcal{S}\rangle$ is $\mathcal{D}(\bar{\mathcal{S}})=\{\{1,2,4\},\{2,3,4,5\},\{4,5,6\}\}$.

Now we are able to formulate the characterization of $\sigma\times n$-degeneracy graphs.

---

[67] Cf. Remark 3.2.2.4 and Def. 3.2.1.3, 3.2.1.7.

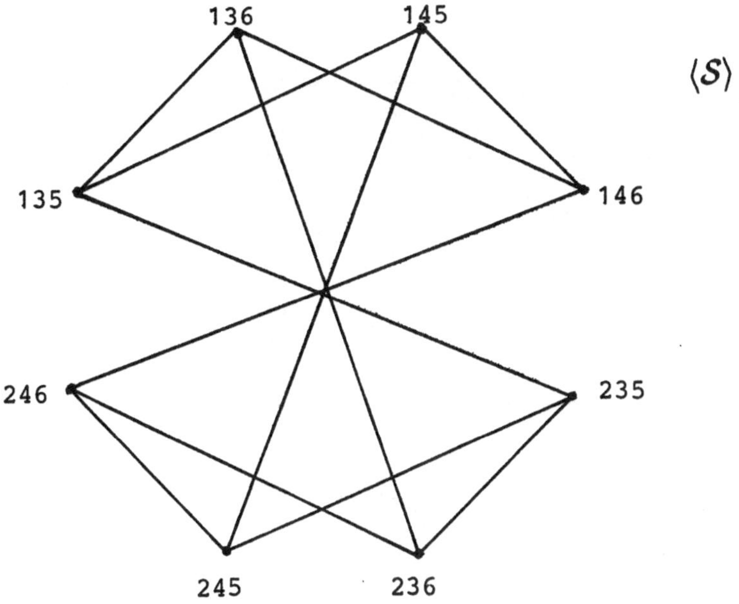

Legend:
i, j, k – node $\{i,j,k\}$ of $\mathcal{S}$

Fig. 3.2.2.2
The 3 × 3-index graph $\langle \mathcal{S} \rangle$ in Example 3.2.2.7a)

Theorem 3.2.2.8:

Let the index graph $\langle \mathcal{S} \rangle$ and its representation system $\mathcal{T} = \mathcal{D}(\bar{\mathcal{S}})$ be given. The graph $\langle \mathcal{S} \rangle$ is a $\sigma \times n$-degeneracy graph[68] if and only if $\mathcal{T}$ is a σ-induced set system on $\{1, \dots, N\}$ $(N = n + \sigma)$. Moreover, it holds that $\langle \mathcal{S} \rangle = G_Y \Leftrightarrow \mathcal{T} = \mathcal{T}_Y$ $(Y \in \mathbb{R}^{\sigma \times n})$.

Proof: In order to prove the theorem, it is enough to prove the last assertion.[69]

a) Let be $\langle \mathcal{S} \rangle = G_Y$. Now $\langle \mathcal{S} \rangle$ consists of the index sets of regular $\sigma \times \sigma$-submatrices of $(Y|I_\sigma)$. Thus the elements of $\bar{\mathcal{S}}$ are index sets of $\sigma \times \sigma$- submatrices of $(Y|I_\sigma)$ with rank $< \sigma$. Finally,

---

[68] Cf. Remark 3.2.2.5. In Section 3.2.2 we assume that the matrix Y of a degeneracy graph G_Y has to be feasibly laid (cf. Kruse (1986:40)).

[69] Cf. Def. 3.1.3.4, 3.2.1.13 and Remark 3.1.3.5.

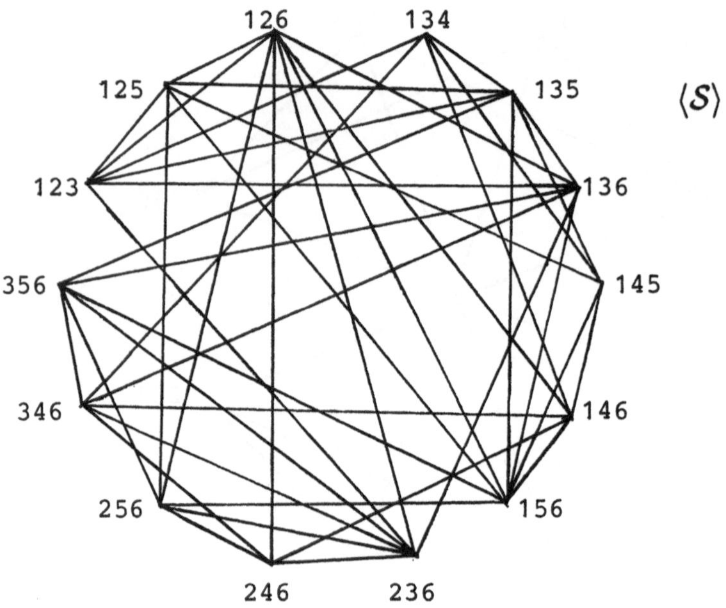

Legend:
$i,\ j,\ k$ – node $\{i,j,k\}$ of $\mathcal{S}$

Fig. 3.2.2.3
The 3 × 3-index graph $\langle \mathcal{S} \rangle$ in Example 3.2.2.7b)

$\mathcal{D}(\bar{\mathcal{S}})$ consists of the index sets of maximal $\sigma \times k$-submatrices of $(Y|I_\sigma)$ with rank $< \sigma$ $(\sigma \leq k \leq N)$, i.e. $\mathcal{T} = \mathcal{T}_Y$.

b) The reverse assertion can be proved analogously.•

The interrelations in the above theorem are illustrated in Fig. 3.2.2.4 which has to be interpreted as follows:

By the mapping $\langle \mathcal{S} \rangle \mapsto \mathcal{T} = \mathcal{D}(\bar{\mathcal{S}})$ the $\sigma \times n$-index graph $\langle \mathcal{S} \rangle$ is biuniquely assigned to a σ-normal set system $\mathcal{T}$. Moreover, the class of $\sigma \times n$-degeneracy graphs is contained in the class of $\sigma \times n$-index graphs, while the class of σ-induced set systems on $\{1,\ldots,N\}$ is contained in the class of σ-normal set systems $(N = n + \sigma)$.

The above characterization asserts that the mapping above assigns $\sigma \times n$-degeneracy graphs biuniquely to σ-induced set systems.

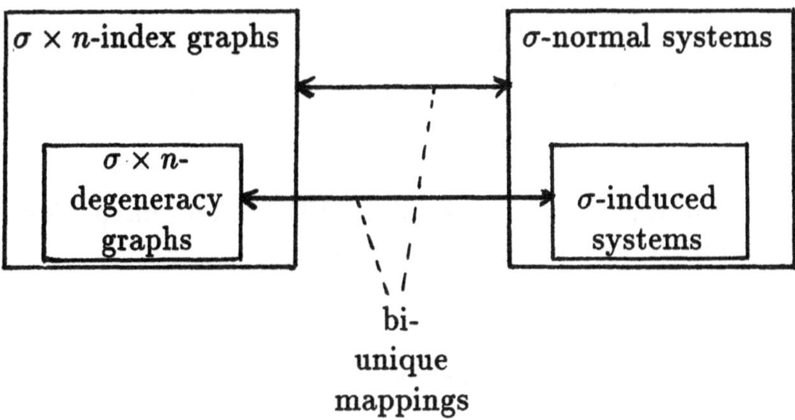

Fig. 3.2.2.4
Conceptual connections in Theorem 3.3.2.8

We will apply Theorem 3.2.2.8 in order to decide whether a given $\sigma \times n$-index graph is a $\sigma \times n$-degeneracy graph[70]. For this purpose we have to prove whether the representation system is σ-induced. The following examples will illustrate this.

Example 3.2.2.9:
Let be $\sigma = 4, n = 2, N = 6$. We consider the 4×2-index graph $\langle \mathcal{S} \rangle$ with node set

$$\mathcal{S} = \{\{1,2,5,6\},\{1,4,5,6\},\{2,3,5,6\},\{3,4,5,6\}\}$$

(cf. Fig. 3.2.2.5). The complementary system is

$$\begin{aligned}\bar{\mathcal{S}} =&\{\{1,2,3,4\},\{1,2,3,5\},\{1,2,3,6\},\{1,2,4,5\}\\ &\{1,2,4,6\},\{1,3,4,5\},\{1,3,4,6\},\{1,3,5,6\}\\ &\{2,3,4,5\},\{2,3,4,6\},\{2,4,5,6\}\}.\end{aligned}$$

Hence the representation system of $\langle \mathcal{S} \rangle$ is

[70] Cf. the introductory representations of Section 3.2.2 on the use of a characterization.

$$\mathcal{T} = \mathcal{D}(\bar{\mathcal{S}}) = \{\{1,2,3,4,5\},\{1,2,3,4,6\},\{2,4,5,6\},\{3,4,5,6\}\}$$

It can be easily proved that $\mathcal{T} = \mathcal{T}_Y$ holds for

$$Y = \begin{pmatrix} 2, & 0 \\ 0, & 2 \\ 0, & 0 \\ 0, & 0 \end{pmatrix} \quad \in I\!R^{4\times 2}$$

i.e. system $\mathcal{T}$ is σ-induced[71] (cf. Def. 3.2.1.13). Theorem 3.2.2.8 implies that the graph $\langle \mathcal{S} \rangle$ is a 4×2-degeneracy graph.

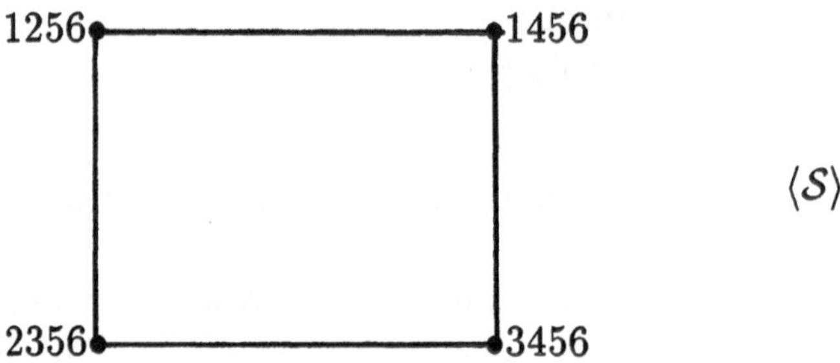

Legend:
i, j, k, l - node $\{i,j,k,l\}$ of $\mathcal{S}$

Fig. 3.2.2.5
The 4×2-degeneracy graph $\langle \mathcal{S} \rangle$ of Example 3.2.2.9

In the example above we construct a matrix Y, in order to decide whether the representation system of a given index graph is σ-induced. It would be desirable to avoid constructing a matrix Y for this purpose. We present a "direct" way to accomplish the task for $\sigma = 3$:

---

[71] If no matrix Y with $\mathcal{T}=\mathcal{T}_Y$ is given, we can try constructing such a matrix by choosing elements of appropriate ($\sigma-$ 1)-dimensional subspaces of $I\!R^{\sigma}$ as column vectors. However, a detailed investigation of these interrelations would be beyond the scope of this publication.

Corollary 3.2.2.10:

Let a $3 \times n$-index graph $\langle \mathcal{S} \rangle$ and its representation system $\mathcal{T} = \mathcal{D}(\bar{\mathcal{S}}) = \{t_i, \ldots, t_q\}$ be given.

The following condition for $\mathcal{T}$ is necessary for $\langle \mathcal{S} \rangle$ to be a $3 \times n$-degeneracy graph:[72]

For all $Q' \subset Q \subset \{1, \ldots, q\}, |Q'| \geq 2$ holds:

$$\cap_{i \in Q} t_i \neq \emptyset \Rightarrow \cap_{i \in Q'} t_i = \cap_{i \in Q} t_i.$$

Proof: The assertion follows immediately from the Theorems 3.2.2.8 and 3.2.1.15.•

Thus for $\sigma = 3$ we can decide on grounds of certain "intersection properties" of the representation system whether a given index graph is a degeneracy graph.

Example 3.2.2.11:

Let be $\sigma = n = 3, N = 6$. We review the index graph $\langle \mathcal{S} \rangle$ in Example 3.2.2.7b) (cf. Fig. 3.2.2.3). The representation system is $\mathcal{T} = \mathcal{D}(\bar{\mathcal{S}}) = \{t_1, t_2, t_3\}$ with $t_1 = \{1, 2, 4\}, t_2 = \{2, 3, 4, 5\}, t_3 = \{4, 5, 6\}$. For $Q' = \{1, 2\}$ and $Q = \{1, 2, 3\}$ holds

$$\cap_{i \in Q} t_i = \{4\} \neq \emptyset$$

and

$$\cap_{i \in Q'} t_i = \{2, 4\} \neq \cap_{i \in Q} t_i = \{4\}.$$

The above corollary implies that $\langle \mathcal{S} \rangle$ is *not* a 3×3-degeneracy graph (cf. Remark 3.2.2.5 and Theorem 3.2.2.8).

3.2.3 PROPERTIES OF $\sigma \times n$-DEGENERACY GRAPHS

In the present section we shall derive some special properties of $\sigma \times n$-degeneracy graphs. The results are partially based on the characteri-

[72] Example 3.2.3.5 demonstrates that this condition does not necessarily hold for the representation systems of $\sigma \times n$-degeneracy graphs with $\sigma > 3$.

zation above (cf. Theorem 3.2.2.8). The following statement is fundamental for the result with respect to the diameter[73] of $\sigma \times n$-degeneracy graphs.

Lemma 3.2.3.1:

Let a $\sigma \times n$-degeneracy graph G_Y be given $(Y \in \mathbb{R}^{\sigma \times n})$[74]. Between any two nodes I , I' of G_Y there exists a path[75] of length $p = |I \backslash I'|$.

The following proof is by induction (cf. Beisel/Mendel (1987:8)):

We assume that the assertion holds for $p = p_0$ with $p_0 \geq 1$. (The case $p = 1$ is trivial, since I,I' are connected by an edge).

Without loss of generality we may further assume $I = \{j_1, \ldots, j_\sigma\}$, $I' = \{j_{p+1}, \ldots, j_\sigma, j_{\sigma+1}, \ldots, j_{\sigma+p}\}$, where $j_1, \ldots, j_{\sigma+p} \in \{1, \ldots, N\}$ denote different indices $(N = n + \sigma)$. Since the column vector of $(Y|I_\sigma)$, associated with $j_{\sigma+1}$, is $\neq 0$ the *Steinitz' exchange theorem*[76] implies the existence of another node (basis) obtained by substituting an appropriate index of I for $j_{\sigma+1}$. Thus, without loss of generality we may assume $I^* = \{j_2, \ldots, j_{\sigma+1}\}$ to be a node of G_Y. Now it holds that $|I' \backslash I^*| = |\{j_{\sigma+2}, \ldots, j_{\sigma+p}\}| = p - 1 = p_0$. On grounds of the first assumption, a path of length p_0 exists between I' and I^*. Since I and I^* are neighbouring nodes, there exists a path of length $p = p_0 + 1$ between I and I'.•

We are now able to state an upper bound for the diameter of degeneracy graphs.[77]

Theorem 3.2.3.2:

Let G_Y be a $\sigma \times n$-degeneracy graph. For the diameter $d = d(G_Y)$ of G_Y holds that

$$d \leq \min\{\sigma, n\}.$$

---

[73] Cf. Appendix, Def. B.6.

[74] Cf. Remarks 3.1.3.5 and 3.1.3.6.

[75] Cf. Appendix, Def. B.4.

[76] Cf. e.g. Kowalsky (1975:37).

[77] Cf. Theorem 3.3.2.1.

Proof: Let I, I' denote any two nodes of G_Y and $p = |I \backslash I'|$. From $|I| = |I'| = \sigma$ it follows that $p \leq \sigma$. On the other hand $p + \sigma \leq n + \sigma$, thus $p \leq n$ and finally $p \leq \min\{\sigma, n\}$. Now Lemma 3.2.3.1 implies that a path of length $p \leq \min\{\sigma, n\}$ exists between I and I'.•

Obviously the diameter of the 4×2-degeneracy graph in Fig. 3.2.2.5 is $d = 2 = \min\{4, 2\}$, i.e. the upper bound in the above theorem is the smallest possible.

The case $d < \min\{\sigma, n\}$ occurs in the following

Example 3.2.3.3:

We chose the 3×3-degeneracy graph G_Y with

$$Y = \begin{pmatrix} 1 & 1 & 1 \\ 1 & 1 & 1 \\ 1 & 1 & 1 \end{pmatrix} \in \mathbb{R}^{3 \times 3}.$$

It is illustrated in Fig. 3.2.3.1. For the diameter d of G_Y it holds that $d = 2 < \min\{3, 3\} = 3$.

Theorem 3.2.3.2 represents an important result in the theory of degeneracy graphs. It means that between any two nodes of a degeneracy graph a path of length$\leq \min\{\sigma, n\}$ exists. Thus a special case of Theorem 3.2.3.2 is the assertion that (general) $\sigma \times n$-degeneracy graphs are always connected (cf. Kruse (1986:28).[78]

The following theorem provides a formula for the number of nodes of arbitrary degeneracy graphs. It makes use of the fact that representation systems are biuniquely assigned to degeneracy graphs (cf. Def. 3.2.2.6).

Theorem 3.2.3.4:

Let a $\sigma \times n$-degeneracy graph $\langle \mathcal{S} \rangle$ and its representation system $\mathcal{T} = \mathcal{D}(\bar{\mathcal{S}}) = \{t_1, \ldots, t_q\}$ be given.[79] Let U denote the number of nodes of

---

[78] Independently from this Jansson (1985:84, Folgerung 6c) proved that general degeneracy graphs are always connected.

[79] Cf. Def. 3.2.2.6.

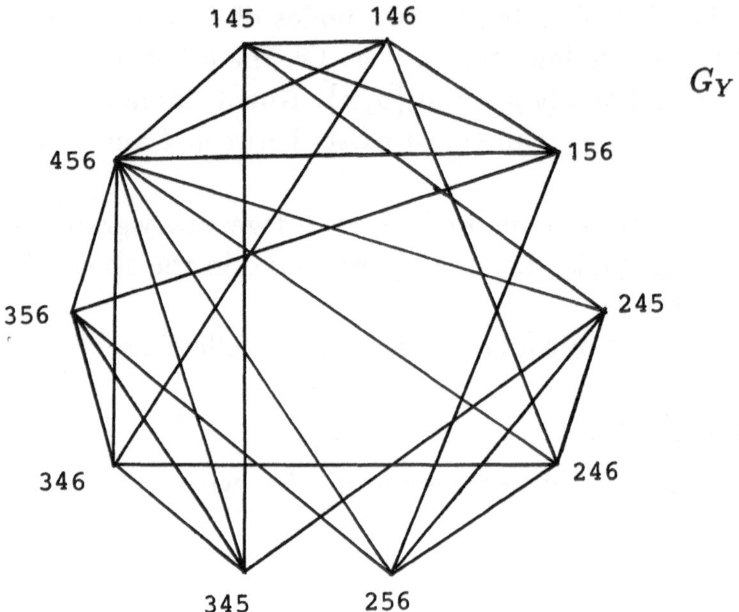

Legend:
$i,\ j,\ k$ – node $\{i,j,k\}$.

Fig. 3.2.3.1
The 3 × 3-degeneracy graph G_Y in Example 3.2.3.3

$\langle\mathcal{S}\rangle$ and $N = \sigma + n, N_{i_1,\ldots,i_r} = |t_{i_1} \cap \ldots \cap t_{i_r}|$ for $1 \leq i_1 < \ldots < i_r \leq q, r \leq q$. Then

$$U = \binom{N}{\sigma} - \sum_{1 \leq i_1 \leq q} \binom{N_{i_1}}{\sigma} + \sum_{1 \leq i_1 < i_2 \leq q} \binom{N_{i_1,i_2}}{\sigma}$$

$$- \sum_{1 \leq i_1 < i_2 < i_3 \leq q} \binom{N_{i_1,i_2,i_3}}{\sigma} + \ldots$$

$$+ (-1)^q \binom{N_{1,\ldots,q}}{\sigma}$$

$$= \sum_{r=0}^{q} (-1)^r \sum_{1 \leq i_1 < \ldots < i_r \leq q} \binom{N_{i_1,\ldots,i_r}}{\sigma},$$

where the term "$r = 0$" in the last sum must be interpreted as $\binom{N}{\sigma}$.

Proof: The nodes of $\langle \mathcal{S} \rangle$ are the σ-subsets of $\{1, \ldots, N\}$ which are contained in none of the sets $t_1, \ldots, t_q$ (cf. Def. 3.2.2.6). If $N'_{i_1,\ldots,i_r}$ $(1 \leq i_1 < \ldots, < i_r \leq q, r \leq q)$ denotes the number of σ-subsets of $\{1, \ldots, N\}$ contained in the intersection of the sets $t_{i_1}, \ldots, t_{i_r}$, the principle of "inclusion and exclusion" of combinatorial analysis[80] implies

$$U = \binom{N}{\sigma} - \sum_{1 \leq i_1 \leq q} N'_{i_1} + \sum_{1 \leq i_1 < i_2 \leq q} N'_{i_1,i_2} - \sum_{1 \leq i_1 < i_2 < i_3 \leq q} N'_{i_1,i_2,i_3} + \ldots + (-1)^q N'_{1,\ldots,q}.$$

Since $N'_{i_1,\ldots,i_r} = \binom{N_{i_1,\ldots,i_r}}{\sigma}$, the assertion is proved.•

Example 3.2.3.5:
We consider the 4×4-degeneracy graph $\langle \mathcal{S} \rangle = G_Y$ with

$$Y = \begin{pmatrix} 0 & 1 & 1 & 0 \\ 2 & 2 & -1 & -1 \\ 0 & 0 & 1 & -1 \\ 0 & 0 & 0 & 0 \end{pmatrix} \in \mathbb{R}^{4 \times 4}$$

which is an induced subgraph of $G^{4,4}$ (cf. Def. 3.2.2.2). The representation system is $\mathcal{T} = \mathcal{T}_Y$ (cf. Theorem 3.2.2.8). Since t_1, t_2, t_3 are the index sets of maximal $4 \times k$-submatrices of $(Y|I_4)$ with rank < 4 $(4 \leq k \leq 8)$, we obtain $\mathcal{T} = \mathcal{T}_Y = \{t_1, t_2, t_3\}$ with

$$t_1 = \{1, 2, 3, 4, 5, 6, 7\},$$
$$t_2 = \{1, 2, 5, 6, 8\},$$
$$t_3 = \{1, 4, 6, 7, 8\},$$

(cf. Def. 3.2.1.11 ; note that the submatrices of $(Y|I_4)$ corresponding to t_1, t_2, t_3 have a zero row and thus have rank < 4).

With the aid of Theorem 3.2.3.4 we draw from $\mathcal{T}$ the number U of nodes of $\langle \mathcal{S} \rangle = G_Y$. We obtain (cf. Fig. 3.2.3.2): $N = 8$, $\sigma = 4$, $q = 3$,

---

[80] Cf. e.g. Aigner (1975:249).

$N_1 = 7,\ N_2 = 5,\ N_3 = 5,\ N_{1,2} = 4,\ N_{1,3} = 4,\ N_{2,3} = 3,\ N_{1,2,3} = 2.$
Thus

$$\begin{aligned} U &= \binom{8}{4} - \binom{7}{4} - \binom{5}{4} - \binom{5}{4} + \binom{4}{4} + \binom{4}{4} + \binom{3}{4} - \binom{2}{4} \\ &= 70 - 35 - 5 - 5 + 1 + 1 + 0 - 0 \\ &= 27. \end{aligned}$$

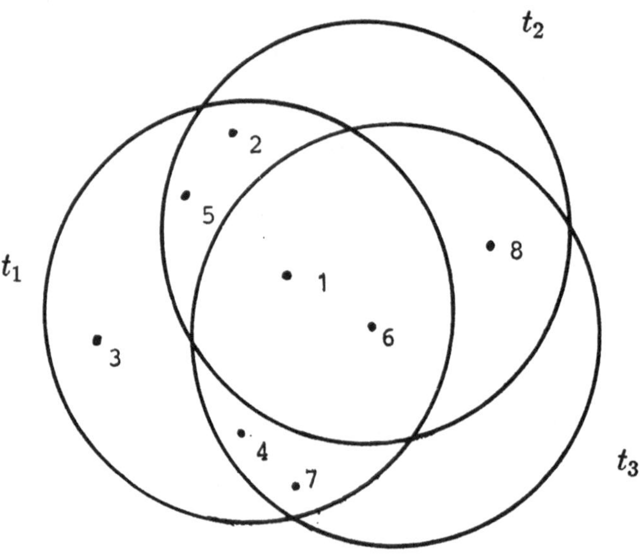

Fig. 3.2.3.2
Illustration of the representation system T in Example 3.2.3.5

As the above example shows, Theorem 3.2.3.4 permits efficient computation of the node number U of degeneracy graphs. Without using this result, expensive computations would be necessary in determining U, since $\binom{8}{4} = 70$ 4×4-submatrices of $(Y\,|\,I_4)$ would have to be checked with regard to regularity.

The following statements refer to the connectivity of (general) degeneracy graphs. We will show that any two nodes of a degeneracy graph can be connected by *at least* two disjoint paths. Thus every degeneracy graph contains at least one cycle (cf. Corollary 3.2.3.9).

Theorem 3.2.3.6:

Let a $\sigma \times n$-degeneracy graph G_Y be given[81]. Any two different nodes I, I' of G_Y can be connected by two *disjoint* paths.

Proof: Without loss of generality we may assume $I = \{j_1, \ldots, j_\sigma\}$ with $1 \leq j_1 < \ldots < j_\sigma \leq n + \sigma - 1$ and $I' = \{n+1, \ldots, n+\sigma\}$. We construct two disjoint paths from I to I' as follows:

a) The *Steinitz' exchange theorem*[82] implies that a further node of G_Y exists, obtained by substituting an appropriate index of I for $n+\sigma$. Thus without loss of generality $I^1 = \{j_1, \ldots, j_{\sigma-1}, n+\sigma\}$ is a node of G_Y (cf. Fig. 3.2.3.3). Moreover, $\tilde{I}^1 = I^1 \backslash \{n+\sigma\}$ and $\tilde{I}' = I' \backslash \{n+\sigma\}$ are nodes of $G_{\tilde{Y}}$, where $\tilde{Y}$ denotes the matrix obtained from Y by omitting the last row. Since every degeneracy graph (and thus G_Y) is connected[83], a path $(\tilde{I}^1, \tilde{I}^2, \ldots, \tilde{I}^k = \tilde{I}')$ exists in $G_{\tilde{Y}}$. Hence $W^1 = (I, I^1, I^2, \ldots, I^k = I')$, where $I^i = \tilde{I}^i \cup \{n+\sigma\}$, ist a path between I and I' in G_Y.

b) Since the last row of Y cannot be a zero row[84], an index $j \in \{1, \ldots, n\}$ exists such that $I_1 = \{j, n+1, \ldots, n+\sigma-1\}$ is a neighbouring node of I'; I and I' are also bases of the tableau

$$\left(\begin{array}{c|cccc} & 1 & 1 & \cdots & 0 \\ & 0 & 1 & \cdots & 0 \\ Y & \vdots & \vdots & & \vdots \\ & 0 & 0 & \cdots & 1 \\ & 0 & 0 & \cdots & 0 \end{array}\right) \in \mathbb{R}^{\sigma \times \sigma+n-1}.$$

Since degeneracy graphs are always connected (cf. part a) of the proof), a sequence of neighbouring nodes $(I_1, I_2, \ldots, I_r = I)$ of

[81] Cf. Remark 3.1.3.6.

[82] Cf. e.g. Kowalsky (1975:37) and the proof of Lemma 3.2.3.1.

[83] Cf. the representations following Example 3.2.3.3.

[84] This would be in contradiction to the above assumption $I \subset \{1, \ldots, n+\sigma-1\}$.

G_Y exists with $I_i \subset \{1,\dots,n+\sigma-1\}$ for $i=1,\dots,r$. From this we obtain a path $W^2 = (I', I_1, I_2, \dots, I_r = I)$ in G_Y, connecting I and I'.

Except for initial and terminal nodes, all nodes of W^1 contain the index $n+\sigma$, while all nodes of W^2 do *not* contain $n+\sigma$. Hence the paths W^1 and W^2 are disjoint, which completes the proof.•

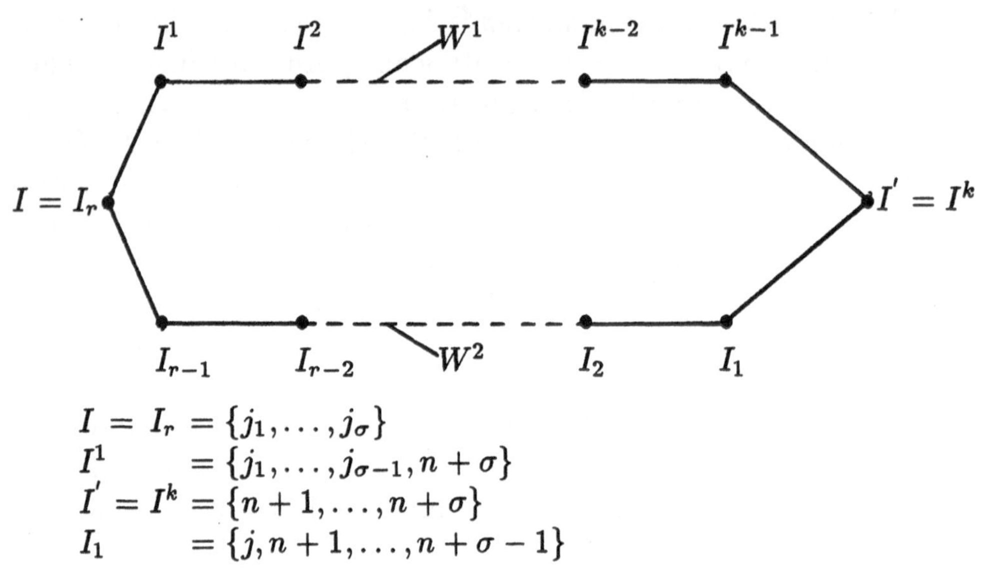

$$\begin{aligned} I = I_r &= \{j_1,\dots,j_\sigma\} \\ I^1 &= \{j_1,\dots,j_{\sigma-1}, n+\sigma\} \\ I' = I^k &= \{n+1,\dots,n+\sigma\} \\ I_1 &= \{j, n+1,\dots,n+\sigma-1\} \end{aligned}$$

Fig. 3.2.3.3
Illustration of the paths W^1, W^2 in G_Y constructed in the proof of Theorem 3.2.3.6

We will illustrate the construction of two disjoint paths in the above proof be means of the following

Example 3.2.3.7:

We consider the degeneracy graph G_Y with

$$Y = \begin{pmatrix} 2 & 0 & 1 \\ 0 & 1 & 0 \\ 0 & 0 & 3 \end{pmatrix} \in \mathbb{R}^{3\times 3} \qquad (\sigma = n = 3).$$

Let the nodes be of the special form $I = \{1,2,3\}$ and $I' = \{4,5,6\}$. The procedure in the proof of Theorem 3.2.3.6 yields the following paths:

a) First of all we determine a neighbouring node I^1 of $I = \{1,2,3\}$ containing the column with index $n + \sigma = 6$. We obtain $I^1 = \{1,2,6\}$ (cf. Fig. 3.2.3.4). Now $\tilde{I}^1 = I^1 \backslash 6 = \{1,2\}$ and $\tilde{I}' = I' \backslash 6 = \{4,5\}$ represent two nodes of $G_{\tilde{Y}}$, where

$$\tilde{Y} = \begin{pmatrix} 2 & 0 & 1 \\ 0 & 1 & 0 \end{pmatrix} \quad \in \mathbb{R}^{2\times 3}.$$

Since $G_{\tilde{Y}}$ is a degeneracy graph and therefore connected, a path between $\tilde{I}^1$ and $\tilde{I}'$ exists in $G_{\tilde{Y}}$, e.g.

$$(\tilde{I}^1 = \{1,2\}, \tilde{I}^2 = \{1,5\}, \tilde{I}^3 = \tilde{I}' = \{4,5\}).$$

From this we obtain the following path between I and I' in G_Y:

$$W^1 = (I = \{1,2,3\}, I^1 = \{1,2,6\}, I^2 = \{1,5,6\}, I^3 = \{4,5,6\} = I').$$

b) The procedure generates the neighbouring node $I_1 = \{3,4,5\}$ of I' (cf. Fig. 3.2.3.4). The nodes I and I_1 are bases of

$$\left(\begin{array}{c|cc} & 1 & 0 \\ Y & 0 & 1 \\ & 0 & 0 \end{array}\right) = \begin{pmatrix} 2 & 0 & 1 & 1 & 0 \\ 0 & 1 & 0 & 0 & 1 \\ 0 & 0 & 3 & 0 & 0 \end{pmatrix}. \qquad (3.2.3.1)$$

Since every degeneracy graph is connected, a sequence of neighbouring bases (regular 3×3-matrices) of (3.2.3.1), connecting the nodes I and I_1, exists; e.g.

$$(I_1 = \{3,4,5\}, I_2 = \{2,3,4\}, I_3 = \{1,2,3\} = I)$$

is such a sequence. This yields the path

$$W^2 = (I' = \{4,5,6\}, I_1 = \{3,4,5\}, I_2 = \{2,3,4\}, I_3 = \{1,2,3\} = I)$$

in G_Y.

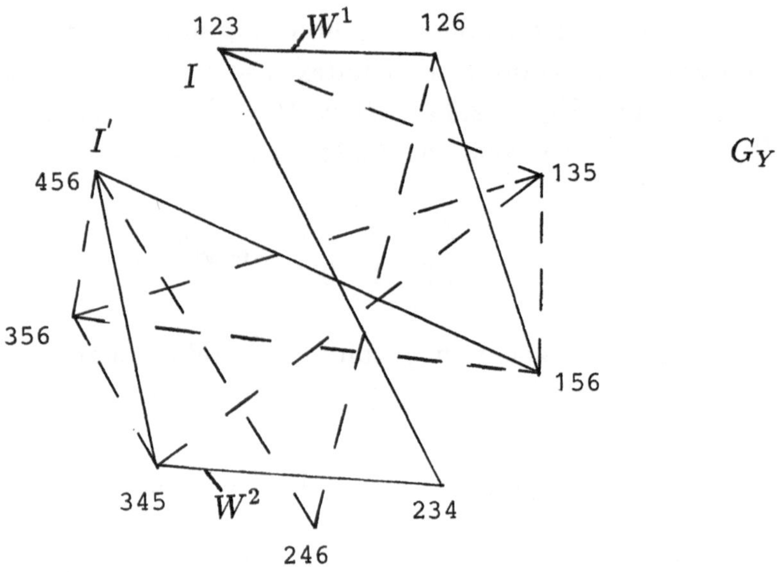

$I = I_3 = \{1,2,3\}; I^1 = \{1,2,6\}; I^2 = \{1,5,6\}; I' = I^3 = \{4,5,6\}; I_1 = \{3,4,5\}; I_2 = \{2,3,4\}$.

Legend:
i, j, k – node of G_Y
•- - -•– edge of G_Y
•——•– edge of W^1 or W^2

Fig. 3.2.3.4
Degeneracy graph G_Y in Example 3.2.3.7 with disjoint paths W^1, W^2 between the nodes $I = \{1,2,3\}$ and $I' = \{4,5,6\}$.

Exept for initial and terminal node, all nodes of W^1 contain index 6, while none of the nodes of W^2 contains this index. Thus the paths W^1 and W^2 are disjoint.

The following results are direct consequences of Theorem 3.2.3.6.

Theorem 3.2.3.8:

The connectivity[85] of any $\sigma \times n$-degeracy graph is ≥ 2.

[85] Cf. Appendix, Def. B.7.

On grounds of numerous examples one could assume that $\sigma \times n$-degeneracy graphs are even n-connected[86]. But further research is necessary to proof this conjecture.

Corollary 3.2.3.9:

Let a $\sigma \times n$-degeneracy graph G_Y be given. For any two nodes I, I' of G_Y exists a cycle in G_Y, containing I and I'.

Proof: We obtain the requested cycle by composing two disjoint paths between I and I' (cf. Theorem 3.2.3.6)[87].•

This result is interesting in connection with explaining simplex cycling, since a "simplex cycle", i.e. the sequence of bases associated with a degenerate vertex, can by represented as a cycle in a (positive) degeneracy graph[88] (cf. Section 4.1).

Though the theory of degeneracy graphs has been developed quite recently (cf. Introduction), diverse applications are conceivable. In particular the above results are useful for the following reasons:

The fact that $s = \min\{\sigma, n\}$ is an upper bound for the diameter of $\sigma \times n$-degeneracy graphs implies e.g. that *every* $\sigma \times n$-degeneracy graph can be theoretically exited or passed in at most s steps (cf. Theorem 3.2.3.2). This is important in connection with developing anticycling rules or testing their efficiency (cf. Section 5.1). Moreover, Theorem 3.2.3.2 could be useful to construct N-trees in solving the neighbourhood problem (cf. Kruse (1986: Section 5.2)), since it insures that each node of a degeneracy graph can be reached in at most s steps, beginning with the starting node.

The formula in Theorem 3.2.3.4 enables us to compute the number of bases (tableaux) associated with a degenerate vertex. Up to now only an upper and a lower bound for this number have been known

---

86 Cf. Appendix, Def. B.7.

87 Cf. Fig. 3.2.3.3.

88 The question, under which conditions certain special cycles exist in graphs or how they can be determined, occurs in different fields of graph theory (cf. Fournier (1985), Fraisse (1986), Jaeger (1985), Richards/Liestman (1985), Volgenant et al.(1986)).

(cf. Kruse (1986:34, 50)). Thus the preceding formula is useful for all possible applications, since the complexity of the basis set decides whether a degeneracy problem is solvable at all.

Finally, sensitivity analysis and determination of shadow prices under degeneracy require information about the connectivity of degeneracy graphs or special subgraphs (optimum graphs).[89]

3.3 THEORY OF $2 \times n$-DEGENERACY GRAPHS

The aim of the present section is to develop an "autonomous" theory for degeneracy graphs with degeneracy degree $\sigma = 2$. The above theory of $\sigma \times n$-degeneracy graphs includes the case $\sigma = 2$, of course, but the following "theory of $2 \times n$-degeneracy graphs" is justified by the fact that their results cannot be obtained from the general statements[90]. It is not possible to generalize the following to the case $\sigma \geq 2$.

3.3.1 CHARACTERIZATION OF $2 \times n$-DEGENERACY GRAPHS

In Section 3.2.2 the $\sigma \times n$-degeneracy graphs with $\sigma, n \in \mathbb{N}$ have been characterized with the aid of certain set systems. In contrast, we can characterize the $2 \times n$-degenerace graphs "directly", i.e. using exclusively graph theoretical concepts.

Theorem 3.3.1.1:

A graph G is a $2 \times n$-degeneracy graph[91] if and only if it is isomorphic to the line graph of any complete r-partite graph[92]

[89] Cf. Piehler/Kruse (1989) and Section 2.3.

[90] E.g. Theorem 3.3.2.5 is essentially "stronger" than Theorem 3.2.3.6 for σ=2. The latter implies that two disjoint paths exist between any two nodes of a $2 \times n$-degeneracy graph. On the other hand Theorem 3.3.2.5 guaranties the existence of *at least* n disjoint connecting paths (cf. also Corollary 3.3.2.8).

[91] In accordance with Kruse (1986:40) we assume in Section 3.3 that the matrix Y of any degeneracy graph G_Y contains neither zero rows nor zero columns (cf. Remarks 3.1.3.5, 3.1.3.6).

[92] Cf. Appendix, Def. B.10, B.12.

$$L(K(p_1,\ldots,p_r))$$

with $r, p_1, \ldots, p_r \in \mathbb{N}$, $r \geq 2$, $p_1 + \ldots + p_r = n + 2$ ($p_1, p_2 > 1$ for $r = 2$).

Proof:

a) Let G_Y be a $2 \times n$-degeneracy graph, where

$$Y = \begin{pmatrix} y_{1,1}, & \ldots, & y_{1,n} \\ y_{2,1}, & \ldots, & y_{2,n} \end{pmatrix} \in \mathbb{R}^{2\times n}$$

contains neither zero rows nor zero columns. We interpret the column vectors of $(Y|I_2)$ as points in the space $\mathbb{R}^2$ (cf. Example 3.3.1.2). Let $\{g_1,\ldots,g_r\}$ denote the minimal system of one-dimensional subspaces of $\mathbb{R}^2$ (= straight lines, passing the origin) with the property that each column vector of $(Y|I_2)$ is contained in exactly one of the spaces $g_1,\ldots,g_r$. Moreover, let p_i denote the number of column vectors, contained in g_i. We now construct a graph by taking the column vectors of $(Y|I_2)$ as nodes and connecting two nodes by an edge if and only if they are contained in different spaces g_i, g_j. By this procedure we obtain the graph $K(p_1,\ldots,p_r)$. Now $G_Y \cong L(K(p_1,\ldots,p_r))$, which results from the following two facts (cf. Remark 3.1.3.6, Def. B.10, B.11):

(1) The nodes of G_Y correspond to the pairs of linear independent column vectors of $(Y|I_2)$ or to the edges of $K(p_1,\ldots,p_r)$ or to the nodes of $L(K(p_1,\ldots,p_r))$.

(2) Two nodes are neighbouring if and only if they differ in exactly one column vector, i.e. if and only if the edges of $K(p_1,\ldots,p_r)$ are neighbouring or if and only if the nodes of $L(K(p_1,\ldots,p_r))$ are neighbouring.
The parameters in $L(K(p_1,\ldots,p_r))$ fulfill the conditions $r \geq 2, p_1+\ldots+p_r = n+2$ ($p_1, p_2 > 1$ for $r = 2$), since Y contains neither zero rows nor zero columns.

b) Conversely, for any line graph of the preceding form we can construct an isomorphic graph of the form G_Y as follows: Let

$\{g_1, \ldots, g_r\}$ denote a system of one-dimensional subspaces[93] of $\mathbb{R}^2$. Now we chose p_i vectors $\neq 0$ from g_i for $i = 1, \ldots, r$. If we conceive them as column vectors, a matrix of the form $(Y|I_2)$ results with $G_Y = L(K(p_1, \ldots, p_r))$.•

The construction of a graph $L(K(p_1, \ldots, p_r))$, isomorphic to a given $2 \times n$-degeneracy graph, is illustrated in

Example 3.3.1.2:

We chose the 2×3-degeneracy graph G_Y with

$$Y = \begin{pmatrix} 2 & 0 & -1 \\ 2 & -2 & -1 \end{pmatrix} \quad \in \mathbb{R}^{2 \times 3}$$

The column vectors $y^1, \ldots, y^5$ of $(Y|I_2)$ – conceived as points in the space $\mathbb{R}^2$ – and the corresponding system of straight lines $\{g_1, g_2, g_3\}$ are illustrated in Fig. 3.3.1.1. If we chose the points $y^1, \ldots, y^5$ as nodes of a graph, which are connected by an edge if and only if they lie on different lines g_i, g_j, the graph $K(1,2,2)$ results. The line graph $L(K(1,2,2))$ is isomorphic to the graph G_Y above (cf. Fig. 3.3.1.2).

Theorem 3.3.1.1 implies the following formula.

Theorem 3.3.1.3:

The number of nonisomorphic $2 \times n$-degeneracy graphs is

$$p(n+2) - 2,$$

where $p(n)$ denotes the number of unordered partitions[94] $n = p_1 + \ldots + p_r$ of the number n $(1 \leq r \leq n)$.

Proof: According to Theorem 3.3.1.1, the partitions $(n+2) = p_1 + \ldots, p_r$ of $(n+2)$ correspond biuniquely to the $2 \times n$-degeneracy graphs of the form $L(K(p_1, \ldots, p_r))$, where the special partitions

$$(n+2) = (n+2) \quad (r = 1)$$

93 The system $\{g_1, \ldots, g_r\}$ must contain an "x- axis" and a "y-axis".

94 Cf. e.g. Hardy/Wright (1958) and Hardy (1959).

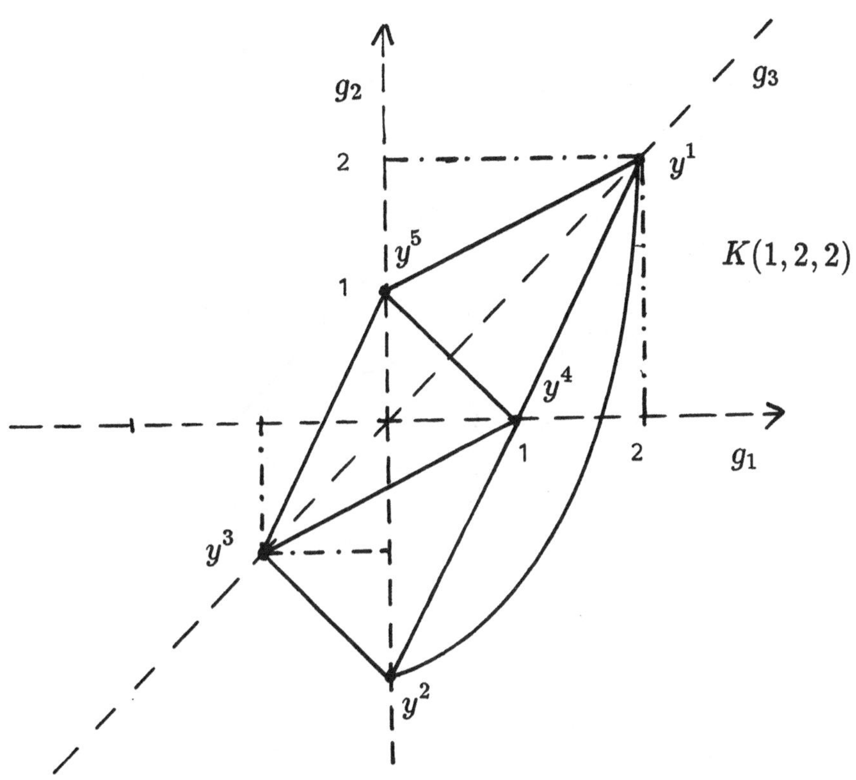

Legend:
y^j - j-th column vector of $(Y \mid I_2)$
g_i - i-th straight line through the origin ($i = 1, 2, 3$)

Fig. 3.3.1.1
Construction of the graph K(1,2,2) in Example 3.3.1.2

and

$$(n+2) = 1 + (n+1) \quad (r = 2, p_1 = 1)$$

of $(n+2)$ are excluded .

Moreover, $K(p_1, \ldots, p_r) \not\cong K(p_1', \ldots, p_q')$ holds for different partitions $(n+2) = p_1 + \ldots + p_r$ and $(n+2) = p_1' + \ldots + p_q'$ of $(n+2)$[95]. From

[95] Note that the degree sequences of $K(p_1, \ldots, p_r)$ and $K(p_1', \ldots, p_q')$ cannot coincide for different partitions (cf. e.g. Behzad/Chartrand (1971:10)).

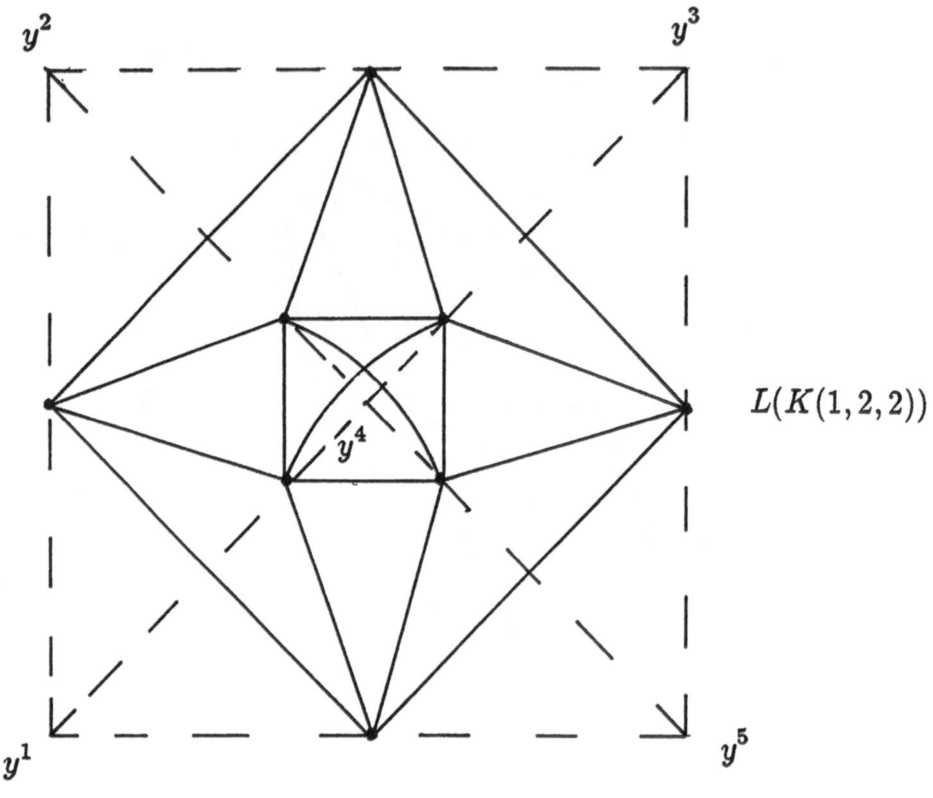

Legend:
y^j – node of $K(1,2,2)$ $(j = 1, \ldots, 5)$
- - - – edge of $K(1,2,2)$
—— – edge of $L(K(1,2,2))$

Fig. 3.3.1.2
Construction of the line graph $L(K(1,2,2)) \cong G_Y$ in Example 3.3.1.2

this follows $L(K(p_1, \ldots, p_r)) \not\cong L(K(p'_1, \ldots, p'_q))$ for different partitions of $(n+2)$, which completes the proof (cf. Behzad/Chartrand (1971:183, Theorem 14.1)).•

Example 3.3.1.4:

Let be $n = 2$. There are $p(4) = 5$ partitions[96] of the number $n + 2 = 4$:

$$\begin{aligned} 4 &= 4 \\ 4 &= 1+3 \\ 4 &= 2+2 \\ 4 &= 1+1+2 \\ 4 &= 1+1+1+1. \end{aligned}$$

Thus $p(4)$-2 $= 3$ nonisomorphic 2×2-degeneracy graphs exist (cf. Tab. 3.3.2.1): $L(K(2,2)), L(K(1,1,2)), L(K(1,1,1,1))$.

3.3.2 PROPERTIES OF $2 \times n$-DEGENERACY GRAPHS

The characterization of $2 \times n$-degeneracy graphs as line graphs of complete r-partite graphs means that the former can be "reduced" to r-partite graphs in a certain sense, e.g. that nodes of a $2 \times n$-degeneracy graph correspond to edges of a complete r-partite graph, paths in $2 \times n$-degeneracy graphs correspond to sequences of neighbouring edges in complete r-partite graphs, etc.

Based on this "reduction principle", the following properties of $2 \times n$-degeneracy graphs are derived.[97]

Theorem 3.3.2.1:

The diameter of every $2 \times n$-degeneracy graph is ≤ 2.

Proof: Let $L(G)$ be a $2 \times n$-degeneracy graph with $G = K(p_1, \ldots, p_r)$. Moreover, let $K = \{v_1, v_2\}$ and $K' = \{v_3, v_4\}$ denote two distinct edges of G or nodes of $L(G)$.

We have to distinguish between the following two cases:

a) The edges K, K' of G have a common node (in G). Now the distance between K and K', interpreted as nodes in $L(G)$, is 1.

---

[96] Extensive tables of partitions are provided in Gupta (1962).

[97] Cf. the introductory representations in Section 3.2.2.

b) The edges K and K' have no common node. Then $v_1, \ldots, v_4$ are distinct nodes. Without loss of generality we may assume that v_1 and v_4 belong to different components of the partition $V = V_1 \cup \ldots \cup V_r$ of V (cf. Appendix, Def. B.10). Now $(\{v_1, v_2\}, \{v_1, v_4\}, \{v_3, v_4\})$ is a path of length two, connecting K and K' in $L(G)$ (cf. Fig. 3.3.2.1). Thus the distance between the nodes K, K' in $L(G)$ is 2.•

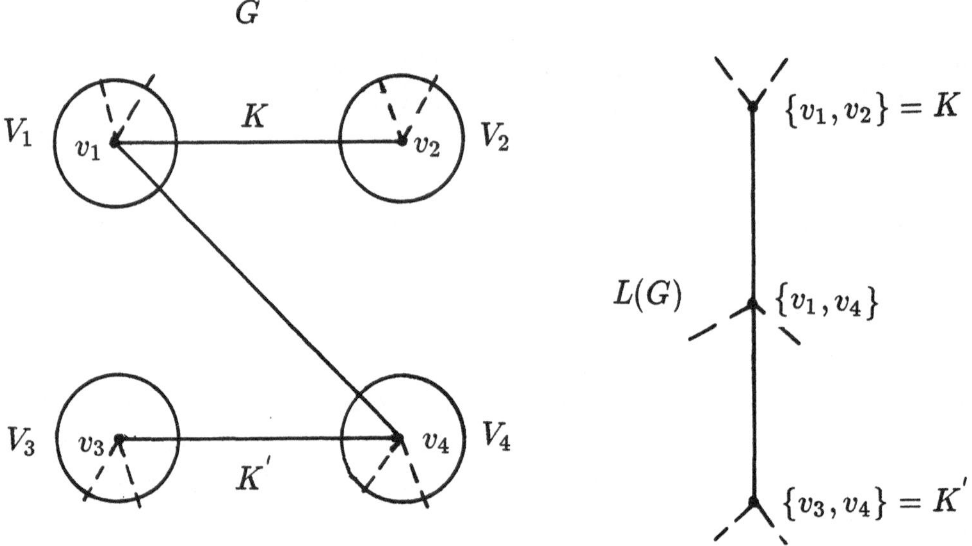

Legend:
v_i – node of G ($i = 1, \ldots, 4$)
V_i – i-th component of the partition of V ($i = 1, \ldots, 4$)
K, K' – edge of G or node of $L(G)$

Fig. 3.3.2.1
Construction of a path in case b) of the proof of Theorem 3.3.2.1

Important structural properties of $2 \times n$-degeneracy graphs are provided in

Theorem 3.3.2.2:

Let $L(G)$ be a $2 \times n$-degeneracy graph with $G = K(p_1, \ldots, p_r)$ ($r \geq 2, p_1 + \ldots + p_r = n+2; p_1, p_2 > 1$ for $r = 2$). Moreover, let U, U' denote the number of nodes or edges of $L(G)$, respectively. Then holds:

a) The node set V^* of $L(G)$ can be partitioned in components $V^*_{i,j}$ such that $V^*_{i,j}$ contains $p_i \cdot p_j$ nodes , all of which have degree $2n + 2 - p_i - p_j$ $(1 \leq i < j \leq r)$.

b) $U = \frac{1}{2}((n+2)^2 - \sum_{i=1}^{r} p_i^2)$

c)[98] $U' = \sum_{i=1}^{r} p_i \binom{n+2-p_i}{2}$

Proof:

a) Let $V = V_1 \cup \ldots \cup V_r$ be the partition of the node set V of G with $|V_i| = p_i$ for $i = 1, \ldots, r$[99]. Let further $V^*_{i,j}$ denote the set of all edges of G, connecting nodes of different components V_i, V_j $(1 \leq i < j \leq r)$ (cf. Fig. 3.3.2.2). Obviously the number of edges in $V^*_{i,j}$ is $p_i \cdot p_j$. Let now $K = \{v_1, v_2\}$ denote any edge in $V^*_{i,j}$ $(v_1 \in V_i, v_2 \in V_j)$. Then K intersects $n + 1 - p_i$ further edges of G in v_1, or $n + 1 - p_j$ edges in v_2. Thus K has $2n + 2 - p_i - p_j$ neighbouring edges in G. Interpreting edges of G as nodes of $L(G)$ yields assertion a).

b) From a) follows

$$U = \sum_{1 \leq i < j \leq r} p_i \cdot p_j$$

$$= \frac{1}{2}((\sum_{i=1}^{r} p_i)^2 - \sum_{i=1}^{r} p_i^2)$$

which implies the formula in b).

98 An equivalent formula for the number of edges is derived in Zörnig (1985:9).

99 Cf. Appendix, Def. B.10.

c) Let $g(v)$ denote the degree of a node v of G, then

$$\begin{aligned} U' &= \sum_{v \in V} \binom{g(v)}{2} \\ &= \sum_{i=1}^{r} \sum_{v \in V_i} \binom{g(v)}{2} \\ &= \sum_{i=1}^{r} p_i \binom{n+2-p_i}{2}. \end{aligned}$$

The last equation follows from $|V_i| = p_i$ and the fact that all nodes in component V_i of V have degree $n + 2 - p_i$.•

Example 3.3.2.3:
We consider the 2×3-degeneracy graph $L(G)$ with $G = K(1,2,2)$ ($p_1 = 1, p_2 = p_3 = 2; r = n = 3$). The above theorem yields the following results, which can be verified easily be means of Fig. 3.3.1.2 (cf. also Tab. 3.3.2.1):

a) $L(G)$ has
$p_1 \cdot p_2 = 2$ nodes with degree $2n + 2 - p_1 - p_2 = 5$,
$p_1 \cdot p_3 = 2$ nodes with degree $2n + 2 - p_1 - p_3 = 5$,
$p_2 \cdot p_3 = 4$ nodes with degree $2n + 2 - p_2 - p_3 = 4$.

b) For the node number U of $L(G)$ holds

$$\begin{aligned} U &= \frac{1}{2}((n+2)^2 - \sum_{i=1}^{r} p_i^2) \\ &= \frac{1}{2}((3+2)^2 - 1^2 - 2^2 - 2^2) = 8. \end{aligned}$$

c) For the edge number U' of $L(G)$ holds

$$\begin{aligned} U' &= \sum_{i=1}^{r} p_i \binom{n+2-p_i}{2} \\ &= 1 \cdot \binom{5-1}{2} + 2 \cdot \binom{5-2}{2} + 2 \cdot \binom{5-2}{2} = 18. \end{aligned}$$

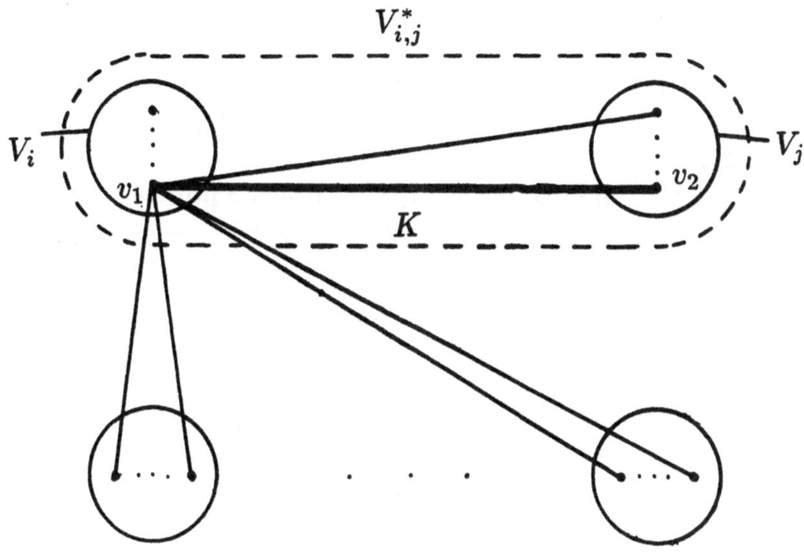

Legend:
v_1, v_2 - node of G
V_i, V_j - Component of the partitition of V

Fig. 3.3.2.2
Representation of an edge $K = \{v_1, v_2\} \in V^*_{i,j}$ and their neighbouring edges, containing v_1 (cf. proof of Theorem 3.3.2.2)

Theorem 3.3.2.2b) can be used in particular to determine the maximal and the minimal node number of a $2 \times n$-degeneracy graph (cf. Zörnig (1985:11) and Kruse (1986:34, 50)).

The following lemma, based on the "reduction principle", is fundamental to determining the connectivities[100] of $2 \times n$-degeneracy graphs.

---

[100] Cf. Appendix, Def. B.7.

Lemma 3.3.2.4:

Let a complete r-partite graph $G = K(p_1, \ldots, p_r)$ and two distinct edges $K = \{v_1, v_2\}$ and $K' = \{v_3, v_4\}$ of G be given ($r \geq 2, p_1 + \ldots + p_r = n + 2$).

Moreover, let $V^{(i)}$ denote the component of the node set- partition[101] $V = V_1 \cup \ldots \cup V_r$, which contains v_i; $k^{(i)} := |V^{(i)}|$ $(i = 1, \ldots, 4)$. Without loss of generality we may assume: $k^{(3)} + k^{(4)} \geq k^{(1)} + k^{(2)}$.

Then $2n+2-k^{(3)}-k^{(4)}$ pairwise disjoint edge paths[102] exist between K and K'.

Proof:[103] We construct the above number of pairwise disjoint edge paths between K and K' for each of the following cases:

Case 1:

$V^{(1)}, \ldots, V^{(4)}$ are distinct components of V

First of all we obtain 4 disjoint edge paths of the form (cf. Fig. 3.3.2.3):

$$W_{v_i, v_j} := (\{v_1, v_2\}, \{v_i, v_j\}, \{v_3, v_4\}) \quad \text{for} \quad i \in \{1; 2\}, j \in \{3; 4\}.$$

Moreover, there are $n - k^{(1)} - k^{(3)}$ edge paths of the form

$$W_{v_1, v, v_3} := (\{v_1, v_2\}, \{v_1, v\}, \{v_3, v\}, \{v_3, v_4\}) \quad \text{for} \quad v \in V \backslash (V^{(1)} \cup V^{(3)} \cup \{v_2, v_4\})$$

and $n - k^{(2)} - k^{(4)}$ edge paths of the form

$$W_{v_2, v', v_4} := (\{v_1, v_2\}, \{v_2, v'\}, \{v_4, v'\}, \{v_3, v_4\}) \quad \text{for} \quad v' \in V \backslash (V^{(2)} \cup V^{(4)} \cup \{v_1, v_3\}).$$

---

101 $V^{(1)}, \ldots, V^{(4)}$ denote not necessarily different components in contrast to $V_1, \ldots, V_r$ (cf. B.10).

102 Cf. Appendix, Def. B.14.

103 An alternative proof is by induction.

Finally we obtain $k^{(1)} + k^{(2)} - 2$ edge paths of the form (cf. Fig. 3.3.2.4)

$$W^{(t)} := (\{v_1, v_2\}, \{v_{f(t)}, w_t'\}, \{w_t, w_t'\}, \{w_t, v_{g(t)}\}, \{v_3, v_4\}) \quad \text{for} \quad t = 1, \ldots, k^{(1)} + k^{(2)} - 2,$$

where $w_1, \ldots, w_{k^{(1)}+k^{(2)}-2}$ and $w_1', \ldots, w_{k^{(3)}+k^{(4)}-2}'$ denote the elements of $(V^{(1)} \cup V^{(2)}) \backslash \{v_1, v_2\}$ or of $(V^{(3)} \cup V^{(4)}) \backslash \{v_3, v_4\}$, respectively and

$$f(t) := \begin{cases} 1 & \text{if} \quad w_t' \in V^{(3)} \\ 2 & \text{if} \quad w_t' \in V^{(4)} \end{cases}$$

$$g(t) := \begin{cases} 3 & \text{if} \quad w_t \in V^{(1)} \\ 4 & \text{if} \quad w_t \in V^{(2)} \end{cases}$$

Together we obtain

$$4+(n-k^{(1)}-k^{(3)})+(n-k^{(2)}-k^{(4)})+(k^{(1)}+k^{(2)}-2) = 2n+2-k^{(3)}-k^{(4)}$$

edge paths between K and K' which are obviously pairwise disjoint.

Case 2:
$V^{(1)} = V^{(3)}; V^{(1)} \neq V^{(2)} \neq V^{(4)}, V^{(1)} \neq V^{(4)}$

a) $v_1 \neq v_3$
First we obtain the edge paths (cf. Fig. 3.3.2.5)

$$W_{v_1,v_4}, W_{v_2,v_3}, W_{v_2,v_4}.$$

Moreover, there are $n - k^{(3)}$ edge paths of the form

$$W_{v_1,v,v_3} \quad \text{for} \quad v \in V \backslash (V^{(3)} \cup \{v_2, v_4\})$$

and $n - k^{(2)} - k^{(4)}$ edge paths of the form

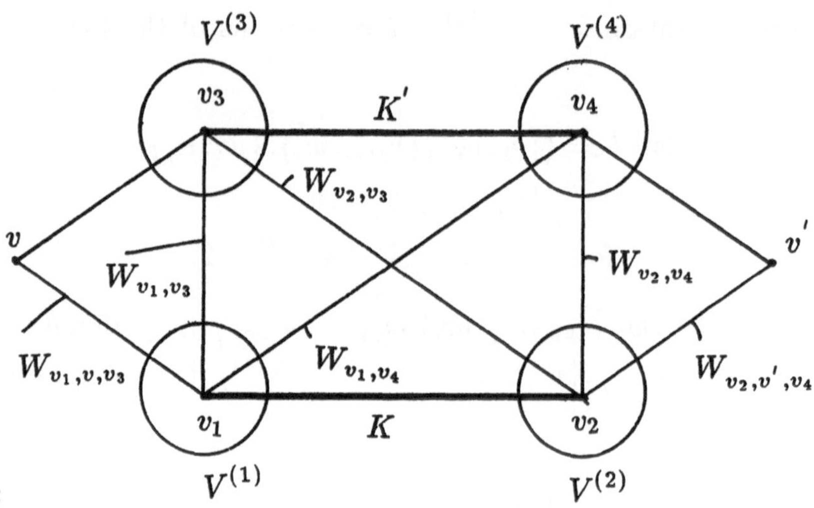

Legend:

v, v', v_i – nodes of G

$V^{(i)}$ – component of the partition of V

K, K' – edges

$W \ldots$ – edge path

Fig. 3.3.2.3

Illustration of edge paths between K and K' in case 1

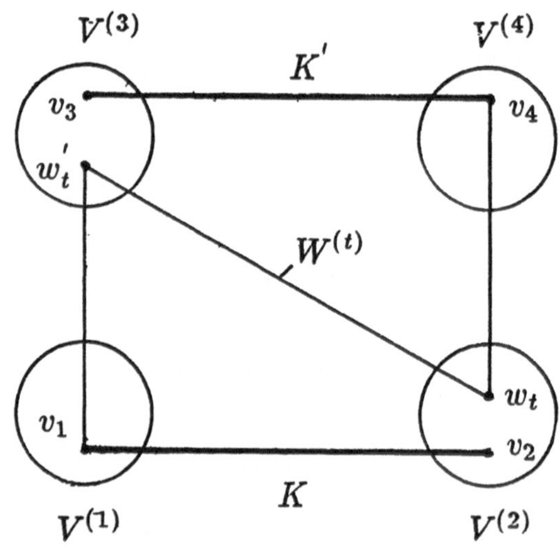

Legend:

v_i, w_t, w_t' – nodes of G

$V^{(i)}$ – component of the partition of V

K, K' – edges

$W^{(t)}$ – edge path

Fig. 3.3.2.4

Illustration of edge paths of the form $W^{(t)}$ in case 1

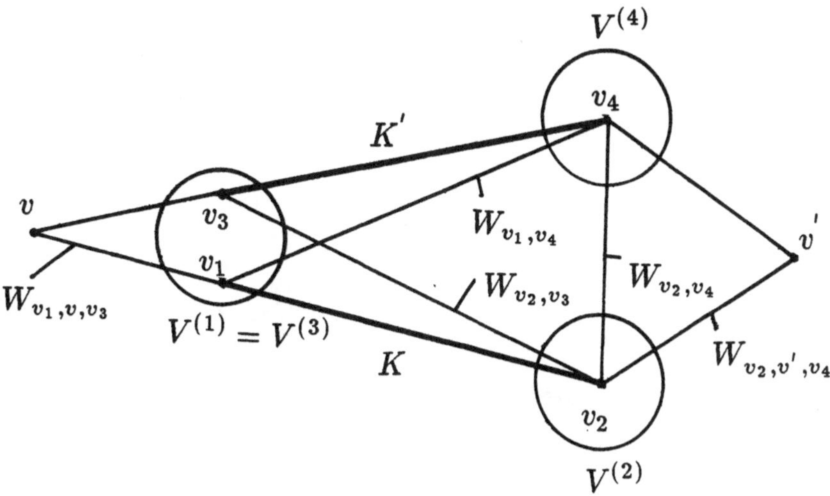

Legend:
v, v', v_i – nodes of G
$V^{(i)}$ – component of the partition of V
K, K' – edges
$W \ldots$ – edge path

Fig. 3.3.2.5
Illustration of edge paths between K and K' in case 2a

$$W_{v_2,v',v_4}$$
$$\text{for} \quad v' \in V \backslash (V^{(2)} \cup V^{(4)} \cup \{v_1, v_3\})$$

(notations as in case 1).

Similarly to case 1 we furthermore obtain $k^{(2)} - 1$ edge paths of the form

$$W'^{(t)} := (\{v_1, v_2\}, \{v_2, w_t'\}, \{w_t, w_t'\}, \{w_t, v_4\}, \{v_3, v_4\})$$
$$\text{for} \quad t = 1, \ldots, k^{(2)} - 1,$$

where $w_1, \ldots, w_{k^{(2)}-1}$ and $w_1', \ldots, w_{k^{(4)}-1}'$ denote the elements of $V^{(2)} \backslash \{v_2\}$ or $V^{(4)} \backslash \{v_4\}$, respectively. Together we obtain

$$3 + (n - k^{(3)}) + (n - k^{(2)} - k^{(4)}) + (k^{(2)} - 1) = 2n + 2 - k^{(3)} - k^{(4)}$$

pairwise disjoint edge paths between K and K'.

b) $v_1 = v_3$

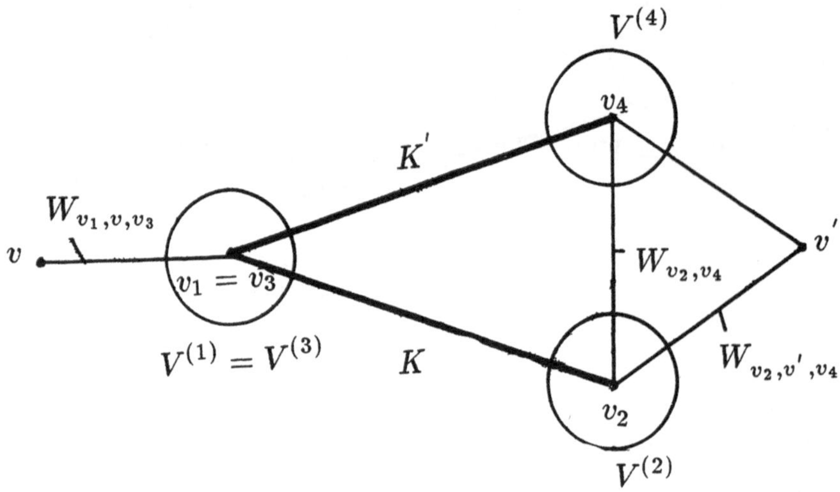

Legend:
v, v', v_i – nodes of G
$V^{(i)}$ – component of the partition of V
K, K' – edges
$W \ldots$ – edge path

Fig. 3.3.2.6
Illustration of edge paths between K and K' in case 2b

We obtain the edge paths

$$(\{v_1, v_2\}, \{v_3, v_4\})$$

and W_{v_2,v_4} (cf. Fig. 3.3.2.6). Moreover, there are $n - k^{(3)}$ edge paths of the form

$$W_{v_1,v,v_3} := (\{v_1, v_2\}, \{v_1, v\}, \{v_3, v_4\})$$
$$\text{for} \quad v \in V \backslash (V^{(3)} \cup \{v_2, v_4\})$$

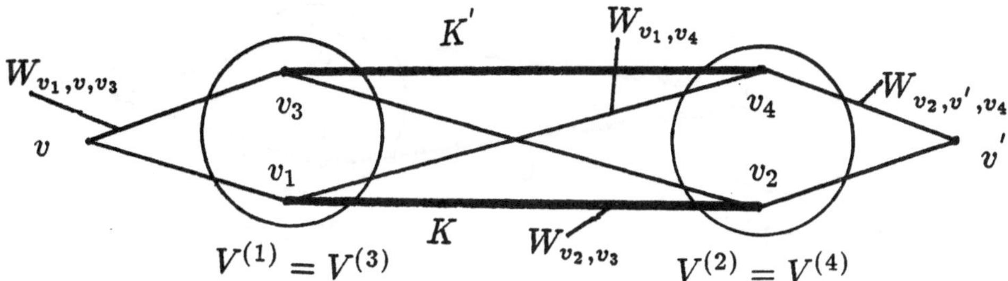

Legend:
v, v', v_i – nodes of G
$V^{(i)}$ – component of the partition of V
K, K' – edges
$W \ldots$ – edge path

Fig. 3.3.2.7
Illustration of edge paths between K and K' in case 3a

and $n+1-k^{(2)}-k^{(4)}$ edge paths of the form

$$W_{v_2,v',v_4}$$
$$\text{for} \quad v' \in V \backslash (V^{(2)} \cup V^{(4)} \cup \{v_1\}).$$

Finally, we obtain $k^{(2)}-1$ edge paths of the form

$$W'^{(t)}$$
$$\text{for} \quad t = 1, \ldots, k^{(2)}-1$$

(cf. case 2a). Together there are

$$2 + (n - k^{(3)}) + (n + 1 - k^{(2)} - k^{(4)}) + (k^{(2)} - 1)$$
$$= 2n + 2 - k^{(3)} - k^{(4)}$$

pairwise disjoint edge paths between K and K'.

Case 3
$V^{(1)} = V^{(3)}; V^{(2)} = V^{(4)}; V^{(1)} \neq V^{(2)}$
a) $v_1 \neq v_3, v_2 \neq v_4$

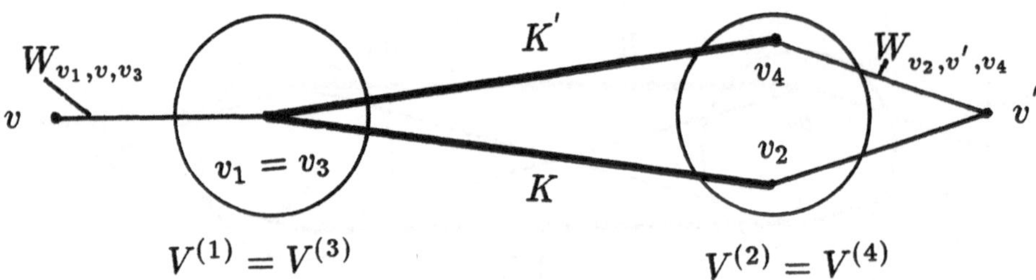

Legend:
v, v', v_i – nodes of G
$V^{(i)}$ – component of the partition of V
K, K' – edges
$W \ldots$ – edge path

Fig. 3.3.2.8
Illustration of edge paths between K and K' in case 3b

We obtain the edge paths (cf. Fig. 3.3.2.7)

$$W_{v_1,v_4} \text{ und } W_{v_2,v_3},$$

$n - k^{(3)}$ edge paths of the form

$$W_{v_1,v,v_3}$$
$$\text{for} \quad v \in V \backslash (V^{(3)} \cup \{v_2, v_4\})$$

and $n - k^{(4)}$ edge paths of the form

$$W_{v_2,v',v_4}$$
$$\text{for} \quad v' \in V \backslash (V^{(2)} \cup \{v_1, v_3\}).$$

Together there are

$$2 + (n - k^{(3)}) + (n - k^{(4)}) = 2n + 2 - k^{(3)} - k^{(4)}$$

pairwise disjoint edge paths between K and K'.

b) $v_1 = v_3, v_2 \neq v_4$

We obtain the edge path (cf. Fig 3.3.2.8)

$$(\{v_1, v_2\}, \{v_3, v_4\}),$$

$n - k^{(3)}$ edge paths of the form (cf. case 2b)

$$W_{v_1, v, v_3}$$
$$\text{for} \quad v \in V \backslash (V^{(3)} \cup \{v_2, v_4\})$$

and $n + 1 - k^{(4)}$ edge paths of the form

$$W_{v_2, v', v_4}$$
$$\text{for} \quad v' \in V \backslash (V^{(2)} \cup \{v_1\}).$$

Together there are

$$1 + (n - k^{(3)}) + (n + 1 - k^{(4)}) = 2n + 2 - k^{(3)} - k^{(4)}$$

pairwise disjoint edge paths between K and K'.•

The above lemma yields the following formulae for the connectivities of $2 \times n$-degeneracy graphs.

Theorem 3.3.2.5:

Let $L(G)$ be a $2 \times n$-degeneracy graph with $G = K(p_1, \ldots, p_r)$ $(p_1 \leq \ldots \leq p_r, r \geq 2, p_1 + \ldots, p_r = n + 2; p_1, p_2 > 1$ for $r = 2)$.

Let ω denote the (node-) connectivity of $L(G)$, then

$$\omega = 2n + 2 - p_{r-1} - p_r.$$

Proof: Lemma 3.3.2.2a implies that $L(G)$ contains at least one node with degree $2n + 2 - p_{r-1} - p_r$. Thus $\omega \leq 2n + 2 - p_{r-1} - p_r$ (cf. Appendix, Def. B.7).

According to Lemma 3.3.2.4, at least $2n+2-p_{r-1}-p_r$ pairwise disjoint paths exist between any two nodes of $L(G)$, since edge paths in G correspond biuniquely to paths in $L(G)$ (recall $p_1 \leq \ldots \leq p_r$).

Thus $\omega \geq 2n+2-p_{r-1}-p_r$.•

Example 3.3.2.6:
We review the 2×3-degeneracy graph $L(G)$ with $G = K(1,2,2)$ ($p_1 = 1, p_2 = p_3 = 2, r = n = 3$). The above theorem yields (cf. Tab. 3.3.2.1)

$$\omega = 2n+2-p_{r-1}-p_r = 2 \cdot 3 + 2 - 2 - 2 = 4.$$

which can be verified easily be means of Fig. 3.3.1.2.

Corollary 3.3.2.7:
Let $L(G)$ be a $2 \times n$-degeneracy graph of the form in Theorem 3.3.2.5.

Let ω' denote the edge connectivity of $L(G)$, then

$$\omega' = \omega = 2n+2-p_{r-1}-p_r$$

Proof: Theorem 3.3.1.5 implies $\omega = 2n+2-p_{r-1}-p_r$. From Theorem 3.3.2.2a follows $\delta = 2n+2-p_{r-1}-p_r$, where δ denotes the minimum degree of $L(G)$. Lemma B.9 completes the proof (cf. Appendix).•

The statements above imply the following bounds for the connectivities.

Corollary 3.3.2.8:
Let $L(G)$ be a degeneracy graph of the form in Theorem 3.3.2.5, then

$$n \leq \omega = \omega' \leq 2n.$$

Proof: The parameters in $L(G)$ obviously fulfill the condition $2 \leq p_{r-1} + p_r \leq n+2$. Now the assertion follows from Corollary 3.3.2.7.•

In graph theory only the line graphs of very special complete r-partite graphs have been investigated so far (cf. Beineke/Wilson (1978, Section 10.5, Theorem 5.6). Thus the above statements on $2 \times n$-degeneracy graphs, which we can conceive as line graphs of *general* complete r-partite graphs (excluding "trivial" exceptions, cf. Theorem 3.3.1.1) are interesting for pure graph theory as well.

Tab. 3.3.2.1
The $2 \times n$-degeneracy graphs $L(K(p_1, \ldots, p_r))$ with node number U, edge number U' and connectivities ω and ω' (n = 1, ..., 5).[104]

n	$p(n+2)-2$	$2 \times n$-degeneracy graphs	U	U'	$\omega = \omega'$
1	1	$L(K(1,1,1))$	3	3	2
2	3	$L(K(2,2))$	4	4	2
		$L(K(1,1,2))$	5	8	3
		$L(K(1,1,1,1))$	6	12	4
3	5	$L(K(2,3))$	6	9	3
		$L(K(1,1,3))$	7	15	4
		$L(K(1,2,2))$	8	18	4
		$L(K(1,1,1,2))$	9	24	5
		$L(K(1,1,1,1,1))$	10	30	6
4	9	$L(K(2,4))$	8	16	4
		$L(K(3,3))$	9	18	4
		$L(K(1,1,4))$	9	24	5
		$L(K(1,2,3))$	11	31	5
		$L(K(2,2,2))$	12	36	6
		$L(K(1,1,1,3))$	12	39	6
		$L(K(1,1,2,2))$	13	44	6
		$L(K(1,1,1,1,2))$	14	54	7
		$L(K(1,1,1,1,1,1))$	15	60	8
5	13	$L(K(2,5))$	10	25	5
		$L(K(3,4))$	12	30	5
		$L(K(1,1,5))$	11	35	6
		$L(K(1,2,4))$	14	47	6
		$L(K(1,3,3))$	15	51	6
		$L(K(2,2,3))$	16	58	7
		$L(K(1,1,1,4))$	15	57	7
		$L(K(1,1,2,3))$	17	68	7
		$L(K(1,2,2,2))$	18	75	8
		$L(K(1,1,1,1,3))$	18	78	8
		$L(K(1,1,1,2,2))$	19	85	8
		$L(K(1,1,1,1,1,2))$	20	95	9
		$L(K(1,1,1,1,1,1,1))$	21	105	10

104 Cf. Theorems 3.3.1.1, 3.3.1.3, .3.3.2.2, 3.3.2.5 and Corollary 3.3.2.8.

The structural properties of $2 \times n$-degeneracy graphs with $n \leq 5$ are summarized in Tab. 3.3.2.1.

SUMMARY OF CHAPTER 3

A source of all kinds of degeneracy problems is the complex structure of the degenerate vertex or of the associated basis set (tableau set) which is caused by the structure of the enlarged matrix of coefficients. The former can be represented appropriately by means of degeneracy graphs. Thus a careful study of these graphs is necessary in order to understand the degeneracy phenomenon.

Degeneracy graphs represent a valuable tool for solving degeneracy problems, since they make deeper understanding of such problems possible (cf. Introduction); e.g. the fact that a (positive) degeneracy graph always contains a so-called N-tree is decisive in solving the neighbourhood problem (cf. Kruse (1986:Section 5.2.2)). The more the properties of degeneracy graphs are studied, the more applications are possible.

For that reason Chapter 3 does not deal with special applications but the general theory of degeneracy graphs, and is consequently of "formal character". We pursue two aims in the main:

1) We want to characterize degeneracy graphs, i.e. we are seeking for criteria which enable us to decide "easily" whether a given graph is a degeneracy graph.
2) Based on this characterization, we want to detect special properties of degeneracy graphs (node number, edge number , connectivity etc.).

In order to characterize degeneracy graphs, we have assigned biuniquely certain set systems (representation systems) to them. The necessary set theoretical concepts have been developed for that particular purpose. It has been shown that a graph is a degeneracy graph if and only if it is an index graph, the representation system of which is σ-induced. On this basis we have found an upper bound for the diameter and a formula for the node number of degeneracy graphs. Moreover, we have shown that (general) degeneracy graphs are always 2-connected. Hence they contain at least one cycle.

Based on investigations by Zörnig (1985) we have evolved an "autonomous" theory for $2 \times n$-degeneracy graphs. This theory gets along without set theoretical concepts and provides even more far- reaching results than the general theory, but cannot be generalized to the case $\sigma > 2$ ($\sigma =$ degeneracy degree). We have characterized $2 \times n$-degeneracy graphs using line graphs, yielding formulae for node and edge numbers and connectivities.

The results of Chapter 3 are of practical interest as well (cf. Section 3.2.3) and can be especially used to explain simplex cycling (cf. Chapter 4).

4. CONCEPTS TO EXPLAIN SIMPLEX CYCLING

Since Dantzig (1951) published the simplex method, a multitude of so-called anticycling rules[105] has been developed in order to exclude simplex cycling. In contrast, the reasons for simplex cycling have been scarcely investigated in literature (cf. Gal (1988:21f) and Gal/Kruse/Zörnig (1988:219f)).[106] Only Gassner (1964), Marshall/Suurballe (1969), Ollmert (1965, 1969) and Yudin/Gol'shtein (1965) have been engaged in this problem. However, they essentially state only *necessary* conditions for cycling in form of minimum numbers of rows and columns in linear programming or transportation problems.

The present chapter deals with the question which structural properties of the (enlarged) matrix of coefficients or of the degeneracy graph of the degenerate vertex are *necessary and sufficient* for simplex cycling. Except for the special case with two constraints and four structural variables this problem has not been investigated up to now.[107]

The following concepts use degeneracy graphs in order to answer the above question (in a specified form; cf. (4.1.7)). All statements of this chapter are based on the fact that a "simplex cycle" (i.e. the sequence of bases of the degenerate vertex, cf. Def. 4.1.4) can be represented as a cycle of the (positive) degeneracy graph.

---

105 Cf. Altman (1964), Avis/Chvátal (1978), Azpeitia/Dickinson (1964), Benichou et al. (1977), Bland (1977), Cameron (1987), Charnes (1952), Cheng (1980), Cirinà (1985, 1989), Dantzig et al. (1955), Fleischmann (1970), Harris (1973), Hattersley/Wilson (1988), Joksch (1965), Magnanti/Orlin (1988), Wolfe (1963).

106 This is a surprising fact, since already Beale (1955:269) remarked:"...linear programmers are still intrigued by cycling and seek an understanding of the basic reasons underlying its occurrence."

107 Cf. Yudin/Gol'shtein (1965) and the footnote to system (4.4.2.9).

4.1 SPECIFICATION OF THE QUESTION

Let the linear optimization problem

$$\left.\begin{array}{lll} \max z = & c^T x & (c, x \in \mathbb{R}^n) \\ \text{s.t.} & Ax \leq b & (A \in \mathbb{R}^{m \times n}, b \in \mathbb{R}^m) \\ & x \geq 0 & \end{array}\right\} \qquad (4.1.1)$$

or the corresponding canonical form

$$\left.\begin{array}{lll} \max z = & \bar{c}^T \bar{x} & (\bar{c} = \binom{c}{0} \in \mathbb{R}^{m+n}, \bar{x} = \binom{x}{u}) \\ \text{s.t.} & & \\ & \bar{A}\bar{x} = b & (\bar{A} = (A|I_m)) \\ & \bar{x} \geq 0 & \\ & (I_m \in \mathbb{R}^{m \times m} & (\textit{unit matrix}) \\ & u \in \mathbb{R}^m & (\textit{vector of slack variables})) \end{array}\right\} \qquad (4.1.2)$$

be given (cf. (3.1.1.1),(3.1.1.3)). Moreover, let x^0 be a σ-degenerate vertex of the solution set

$$X = \{x \in \mathbb{R}^n | Ax \leq b, x \geq 0\} \qquad (4.1.3)$$

of (4.1.1) (cf. (3.1.1.2)). A simplex tableau of x^0 can be represented in the form of Tab. 4.1.1.[108]

Tab. 4.1.1

Simplex tableau of a basis of the σ-degenerate vertex x^0

basic indices	$1,\ldots,\sigma$	$\sigma+1,\ldots,m$	$m+1,\ldots,m+n$	
1 $\vdots$ σ	I_σ	$0_{\sigma\times(m-\sigma)}$	$Y = Y_{\sigma\times n}$	$0 \in \mathbb{R}^\sigma$
$\sigma+1$ $\vdots$ m	$0_{(m-\sigma)\times\sigma}$	$I_{m-\sigma}$	$Y_{(m-\sigma)\times n}$	$y \in \mathbb{R}^{m-\sigma}$
Δz_j	$0^T \in \mathbb{R}^\sigma$	$0^T \in \mathbb{R}^{m-\sigma}$	$-d^T \in \mathbb{R}^n$	$z \in \mathbb{R}$

---

108 A simplex tableau differs from the corresponding pivot tableau in the row of relative cost coefficients (cf. Tab. 3.1.3.2).

Definition 4.1.1:

Let the linear optimization problem (4.1.1) or (4.1.2), a σ-degenerate vertex of the solution set X in (4.1.3) and the simplex tableau of x^0 in Tab. 4.1.1 be given.

Then the following linear problem, defined by the subtableau of Tab. 4.1.1 in bold type, is called the *reduced (linear optimization) problem*[109] of (4.1.1) or of (4.1.2) (relative to x^0).

$$\left.\begin{array}{ll} \max z = & \bar{d}^T \bar{x}' \; (\bar{d} = \binom{d}{0} \in \mathbb{R}^{n+\sigma}, \bar{x}' = \binom{x}{u'}) \\ \text{s.t.} & \\ & \bar{Y}\bar{x}' = 0 \; (\bar{Y} = (Y|I_\sigma)) \\ & \bar{x}' \geq 0 \\ & (I_\sigma \in \mathbb{R}^{\sigma \times \sigma} (\textit{unit matrix}), \\ & u' \in \mathbb{R}^\sigma (\textit{vector of slack variables})) \end{array}\right\} \quad (4.1.4)$$

For each basis B of a tableau in the form of Tab. 4.1.1 let B' denote the basis of the respective subtableau in bold type. Then the mapping $B \mapsto B'$ assigns the bases of x^0 biuniquely to the bases of (4.1.4) (cf. Kruse (1986:29)).

Let us assume that cycling occurs when the simplex algorithm is applied to (4.1.1), i.e. the bases $B_1, \ldots, B_k, B_1 \ldots$ of x^0 are generated. Then the application of the simplex algorithm to (4.1.4) yields the corresponding sequence $B_1', \ldots, B_k', B_1' \ldots$. Hence in the explanation of simplex cycling we can confine ourselves to reduced linear optimization problems (i.e. the right-hand side equals zero).

The following cycling example of Beale (1955) (cf. also Dantzig (1966:264)) will illustrate the facts.

Example 4.1.2

Let the linear problem (4.1.2) have the special form

---

109 The system of constraints of (4.1.4) corresponds to the canonical system of x^0 (cf. Def. 3.1.3.3).

$$\left.\begin{array}{lllllllll}\max z = & \frac{3}{4}x_1 & -150x_2 & +\frac{1}{50}x_3 & -6x_4 & & & & \\ \text{s.t.} & & & & & & & & \\ & \frac{1}{4}x_1 & -60x_2 & -\frac{1}{25}x_3 & +9x_4 & +x_5 & & & =0 \\ & \frac{1}{2}x_1 & -90x_2 & -\frac{1}{50}x_3 & +3x_4 & & +x_6 & & =0 \\ & & & x_3 & & & & +x_7 & =1 \\ & & & & & & & x_1,\dots,x_7 & \geq 0 \end{array}\right\} \quad (4.1.5)$$

Then the reduced linear optimization problem of (4.1.5), relative to $x^0 = 0 \in \mathbb{R}^4$, has the form

$$\left.\begin{array}{llllllll}\max z = & \frac{3}{4}x_1 & -150x_2 & +\frac{1}{50}x_3 & -6x_4 & & & \\ \text{s.t.} & & & & & & & \\ & \frac{1}{4}x_1 & -60x_2 & -\frac{1}{25}x_3 & +9x_4 & +x_5 & & =0 \\ & \frac{1}{2}x_1 & -90x_2 & -\frac{1}{50}x_3 & +3x_4 & & +x_6 & =0 \\ & & & & & & x_1,\dots,x_6 & \geq 0 \end{array}\right\} \quad (4.1.6)$$

The application of the simplex algorithm to (4.1.5) yields the sequence of tableaux in Tab. 4.1.2, associated with the degenerate vertex $x^0 = 0 \in \mathbb{R}^4$. The subtableaux in bold type represent the corresponding cycle of (4.1.6).

If the simplex algorithm cycles in the application to a reduced problem (4.1.4), the tableau sequence can be represented as a cycle of the degeneracy graph G_Y^+.[110] On the other hand, not every cycle of G_Y^+ represents necessarily the tableau sequence of a cycling example.

Example 4.1.3:

Consider the reduced linear optimization problem (4.1.6) in Example 4.1.2. The tableau sequence of the cycling example in Tab. 4.1.2 is represented by the cycle $C = (\{1,2\},\{2,3\},\{3,4\},\{4,5\},\{5,6\},\{1,6\}, \{1,2,\})$ in Fig. 4.1.1; e.g. the cycle $C' = (\{1,2\},\{2,6\},\{3,6\},\{1,3\}, \{1,2\})$ of G_Y^+ does *not* represent the tableau sequence of a cycling example. (Note that x_6 cannot enter the basis associated with the third tableau in Tab. 4.1.2).

---

[110] Cf. Def. 3.1.3.4, Remarks 3.1.3.5, 3.1.3.6 and Kruse (1984a:17).

Tab. 4.1.2
Tableau sequence of Example 4.1.2 for simplex cycling

	1	2	3	4	5	6	7	x_B
5	($\frac{1}{4}$)	-60	$-\frac{1}{25}$	9	1	0	0	0
6	$\frac{1}{2}$	-90	$-\frac{1}{50}$	3	0	1	0	0
7	0	0	1	0	0	0	1	1
Δz_j	$-\frac{3}{4}$	150	$-\frac{1}{50}$	6	0	0	0	0
1	1	-240	$-\frac{4}{25}$	36	4	0	0	0
6	0	(30)	$\frac{3}{50}$	-15	-2	1	0	0
7	0	0	1	0	0	0	1	1
Δz_j	0	-30	$-\frac{7}{50}$	33	3	0	0	0
1	1	0	($\frac{8}{25}$)	-84	-12	8	0	0
2	0	1	$\frac{1}{500}$	$-\frac{1}{2}$	$-\frac{1}{15}$	$\frac{1}{30}$	0	0
7	0	0	1	0	0	0	1	1
Δz_j	0	0	$-\frac{2}{25}$	18	1	1	0	0
3	$\frac{25}{8}$	0	1	$-\frac{525}{2}$	$-\frac{75}{2}$	25	0	0
2	$-\frac{1}{160}$	1	0	($\frac{1}{40}$)	$\frac{1}{120}$	$-\frac{1}{60}$	0	0
7	$-\frac{25}{8}$	0	0	$\frac{525}{2}$	$\frac{75}{2}$	-25	1	1
Δz_j	$\frac{1}{4}$	0	0	-3	-2	3	0	0
3	$-\frac{125}{2}$	10500	1	0	(50)	-150	0	0
4	$-\frac{1}{4}$	40	0	1	$\frac{1}{3}$	$-\frac{2}{3}$	0	0
7	$\frac{125}{2}$	-10500	0	0	-50	150	1	1
Δz_j	$-\frac{1}{2}$	120	0	0	-1	1	0	0
5	$-\frac{5}{4}$	210	$\frac{1}{50}$	0	1	3	0	0
4	$\frac{1}{6}$	-30	$-\frac{1}{150}$	1	0	($\frac{1}{3}$)	0	0
7	0	0	1	0	0	0	1	1
Δz_j	$-\frac{7}{4}$	330	$\frac{1}{50}$	0	0	-2	0	0

The next pivot step generates the initial tableau.

Legend: tableau associated with basis B

	column indices	x_B
indices of basis B	$B^{-1}\bar{A}$	$x_B = B^{-1}b$
Δz_j	$c_B^T B^{-1}\bar{A} - \bar{c}$	$z = c_B^T x_B$

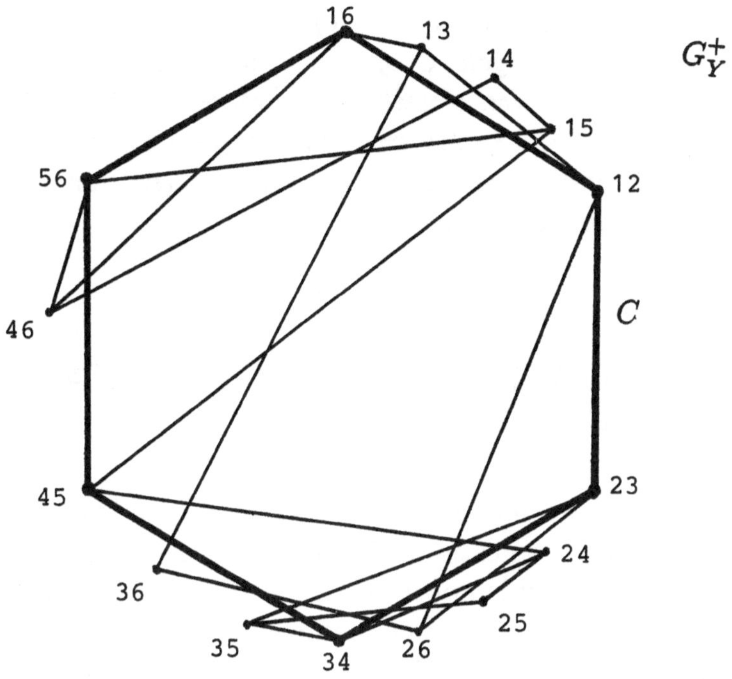

Legend:
i, j – node $\{i,j\}(\{i,j\} \subset \{1,\ldots,5\})$
– edge of G_Y^+
– edge of C

Fig. 4.1.1
Representation of the tableau sequence in Example 4.1.2 as a cycle C in G_Y^+

Definition 4.1.4:

Let the reduced linear optimization problem (4.1.4) be given. A cycle C of the positive degeneracy graph G_Y^+ is called a *simplex cycle* of (4.1.4), if it represents the basis- (tableau-) sequence of a cycling example which can occur in the application of the simplex algorithm to (4.1.4).[111]

We are now able to specify the above question, under which conditions simplex cycling may occur:

---

[111] Obviously, the definition depends on the underlying pivoting rules (cf. Sections 4.2.2 and 4.3.1).

$$\left.\begin{array}{l}\textit{Given the reduced linear optimization problem}\\ \textit{(4.1.4) and a cycle } C \textit{ of the positive degeneracy}\\ \textit{graph } G_Y^+\textit{. Which conditions characterize}^{112} \textit{ sim-}\\ \textit{plex cycles, i.e. which structural properties of } G_Y^+\\ \textit{or of } Y \textit{ and } d\textit{, are necessary and sufficient for } C\\ \textit{to be a simplex cycle?}\end{array}\right\} \qquad (4.1.7)$$

4.2 A PURE GRAPH THEORETICAL APPROACH

In this section we use exclusivly graph theoretical tools in order to answer the above question.[113] Whether the cycle C in (4.1.7) is a simplex cycle, depends not only on the structure of G_Y^+ but also on the vector d of objective function coefficients in (4.1.4).[114]

Hence, the concept of the degeneracy graph has to be modified in so far as the objective function coefficients have to be considered in its definition (cf. Def. 3.1.3.1 and 4.2.1.1).

4.2.1 THE CONCEPT OF THE LP-DEGENERACY GRAPH

For the following definition we modify the initial tableau of (4.1.4) by substitution of the column vector $(0,\ldots,0,1)^T \in \mathbb{R}^{\sigma+1}$ for the right-hand side:

---

112 Cf. the representations of the beginning of Section 3.2.2 on the characterization of graphs.

113 E.g. in Section 4.3 geometrical considerations will also be necessary for this purpose.

114 If we chose e.g. $d_1=0$ for the first coefficient of the objective function in (4.1.6) (where the other coefficients remain unchanged), the cycle C in Fig. 4.1.1 is not a simplex cycle (cf. Example 4.1.3).

Tab. 4.2.1.1
Modified initial tableau[115] of (4.1.4)

basic indices	1	...	n	$n+1$	...	$n+\sigma,$	$n+\sigma+1$
$n+1$	$y_{1,1}$	...	$y_{1,n}$	1	...	0	0
$\vdots$	$\vdots$		$\vdots$	$\vdots$		$\vdots$	$\vdots$
$\vdots$	$y_{\sigma,1}$	...	$y_{\sigma,n}$	0	...	1	0
$n+\sigma+1$	$-d_1$	...	$-d_n$	0	...	0	1

Definition 4.2.1.1:

Let the reduced linear optimization problem (4.1.4) and the modified initial tableau in Tab. 4.2.1.1 be given.

The graph $G_{Y,d} = (V_{Y,d}, E_{Y,d})$ with
$V_{Y,d} = \{\bar{B}|\bar{B}$ is a regular$(\sigma+1)\times(\sigma+1)-$submatrix of Tab. 4.2.1.1$\}$,

$$E_{Y,d} = \{\{\bar{B}, \bar{B}^*\} \subset V_{Y,d}|\bar{B} \leftrightarrow \bar{B}^*\}$$

is called the *(general) LP-degeneracy graph*[116] of (4.1.4).

Analogously we define the *positive (negative) LP-degeneracy graph* $G^+_{Y,d}(G^-_{Y,d})$ if we replace the condition $\bar{B} \leftrightarrow \bar{B}^*$ in $E_{Y,d}$ by $\bar{B} \overset{+}{\longleftrightarrow} \bar{B}^*$ $(\bar{B} \overset{-}{\longleftrightarrow} \bar{B}^*)$.

Remark 4.2.1.2:

a) In the following sense the LP-degeneracy graphs $G_{Y,d}(G^+_{Y,d}, G^-_{Y,d})$ represent enlargements of the (canonical) degeneracy graphs $G_Y(G^+_Y, G^-_Y)$ (cf. Def. 3.1.3.4, Remarks 3.1.3.5, 3.1.3.6): The mapping $\{j_1,\ldots,j_\sigma\} \mapsto \{j_1,\ldots,j_\sigma, n+\sigma+1\}$ $(\{j_1,\ldots,j_\sigma\} \subset \{1,\ldots,n+\sigma\})$ assigns the nodes of $G_Y(G^+_Y, G^-_Y)$ biuniquely to that nodes of $G_{Y,d}(G^+_{Y,d}, G^-_{Y,d})$, which contain the last column of Tab. 4.2.1.1. Moreover, two nodes of $G_Y(G^+_Y, G^-_Y)$ are neighbouring nodes if and only if the corresponding nodes

---

[115] The last row should not be interpreted as the relative cost row but as a "normal" row of the pivot tableau.

[116] The abbreviation "LP" stands for "linear programming".

of $G_{Y,d}(G^+_{Y,d}, G^-_{Y,d})$ are also neighbouring. Hence the graph $G_Y(G^+_Y, G^-_Y)$ is an induced subgraph[117] of $G_{Y,d}(G^+_{Y,d}, G^-_{Y,d})$.

b) We identify a node of an LP-degeneracy graph ($(\sigma + 1) \times (\sigma + 1)$-submatrix of Tab. 4.2.1.1) with its index set (cf. Remark 3.1.3.6).

The definition of the LP-degeneracy graphs is illustrated in

Example 4.2.1.3:
Let the reduced problem (4.1.4) be of the special form

$$\left.\begin{array}{llllll} \max z = & 3x_1 & +2x_2 & & & \\ \text{s.t.} & & & & & \\ & -2x_1 & +x_2 & +x_3 & & = 0 \\ & x_1 & -x_2 & & +x_4 & = 0 \\ & & & & x_1, \ldots, x_4 & \geq 0 \end{array}\right\} \qquad (4.2.1.1)$$

Then the modified initial tableau in Tab. 4.2.1.1 has the form

basic indices	1	2	3	4	5
3	-2	1	1	0	0
4	1	-1	0	1	0
5	-3	-2	0	0	1

Fig. 4.2.1.1 illustrates the LP-degeneracy graphs associated with (4.2.1.1).

4.2.2 CHARACTERIZATION OF SIMPLEX CYCLES BY MEANS OF THE LP-DEGENERACY GRAPH

In the following we always apply the simplex algorithm to the reduced problem (4.1.4). In Section 4.2.2 we assume that *any column* of the simplex tableau *with a negative relative cost coefficient* can be chosen

[117] Cf. Appendix, Def. B.3.

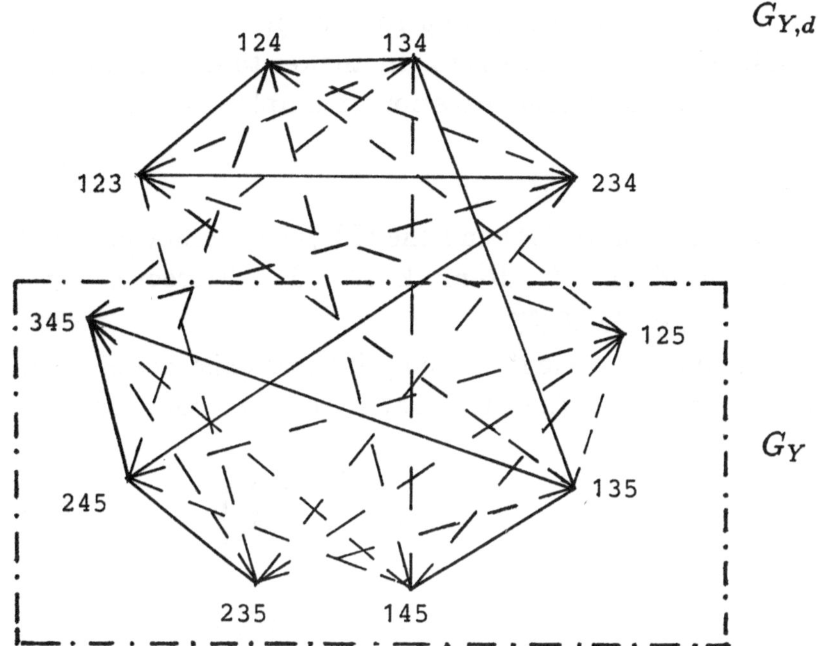

Legend:
$i\ j\ k$ – node[118] $\{i, j, k\}$ of $G_{Y,d}(G^+_{Y,d}, G^-_{Y,d})$
– edge of $G^+_{Y,d}$
– edge of $G^-_{Y,d}$

Fig. 4.2.1.1
The LP-degeneracy graphs $G_{Y,d}(G^+_{Y,d}, G^-_{Y,d})$ and their induced subgraphs $G_Y(G^+_Y, G^-_Y)$ in Example 4.2.1.3

as the pivot column[119] (i.e. the condition $\Delta z_j \leq \Delta z_{j'}$ for all $j' \in \{1, \ldots, n+\sigma\}$ need *not* be satisfied, where j denotes the index of the entering basic variable).

The pivot row is chosen according to the feasibility criterion (cf.

---

[118] Cf. Remark 4.2.1.2.b).
[119] Cf. the introductory representations of Section 4.3.1.

Gal (1987:71)), i.e. we can chose *any positive* pivot element, since the right-hand side of (4.1.4) equals zero. In answer to question (4.1.7) we will now state an initial characterization of simplex cycles:

Theorem 4.2.2.1:

Let the reduced problem (4.1.4) and the (LP)-degeneracy graphs $G_{Y,d}$, $G^+_{Y,d}, G^-_{Y,d}, G^+_Y$ be given (cf. Remark 4.2.1.2). Moreover, let $C = (I_1, \ldots, I_k, I_1)$ be a cycle of G^+_Y.

Then C is a simplex cycle if and only if there exist nodes $\bar{I}_1, \ldots, \bar{I}_k$ of $G_{Y,d}$ (which are *not* nodes of G^+_Y) satisfying the conditions

$$\left.\begin{array}{lll} \bar{I}_\nu \overset{-}{\longleftrightarrow} I_\nu & \text{for} & \nu = 1, \ldots, k \\ \bar{I}_\nu \overset{+}{\longleftrightarrow} I_{\nu+1} & \text{for} & \nu = 1, \ldots, k \\ (I_{k+1} := I_1) & & \end{array}\right\} \tag{4.2.2.1}$$

i.e. if and only if C can be enlarged to a "star-shaped" graph S_C contained in $G_{Y,d}$ (cf. Fig. 4.2.2.1).

Proof:

a) Let C be a simplex cycle. Without loss of generality we may assume for any fixed $\nu \in \{1, \ldots, k\}$ that

$$I_\nu = \{1, \ldots, \sigma, n + \sigma + 1\},$$

$$I_{\nu+1} = \{1, \ldots, \sigma - 1, \sigma + 1, n + \sigma + 1\}.$$

Let the node $\bar{I}_\nu$ of $G_{Y,d}$ be defined by

$$\bar{I}_\nu = (I_\nu \cup I_{\nu+1}) \backslash \{n + \sigma + 1\} = \{1, \ldots, \sigma + 1\}.$$

We will now to prove that the following relations exist between the above nodes:

$$\bar{I}_\nu \overset{-}{\longleftrightarrow} I_\nu \tag{4.2.2.2}$$

$$\bar{I}_\nu \overset{+}{\longleftrightarrow} I_{\nu+1} \tag{4.2.2.3}$$

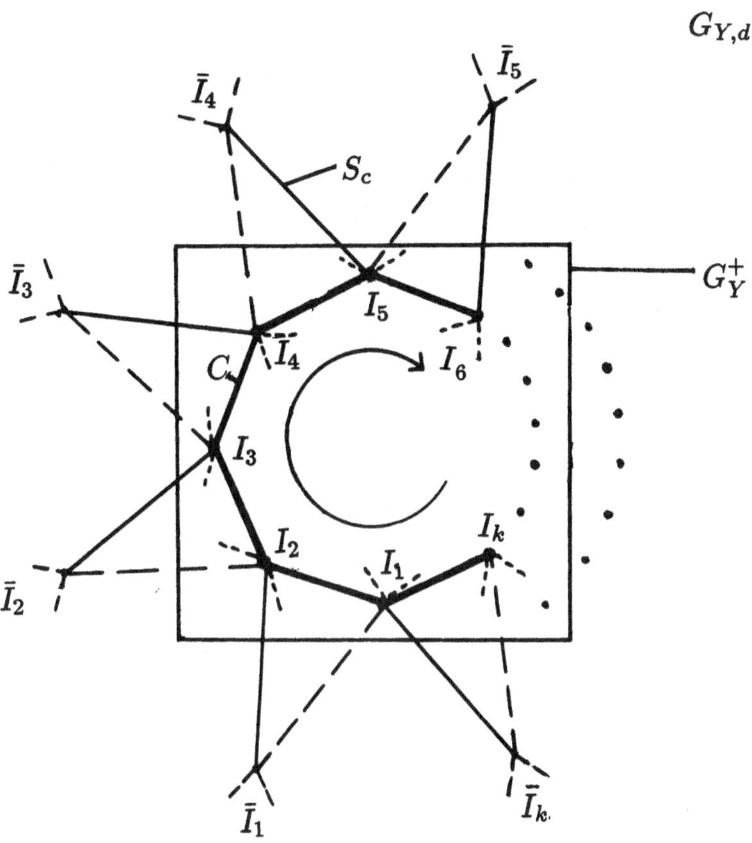

Legend:
I_ν – node of G_Y^+
$\bar{I}_\nu$ – node of $G_{Y,d}$
(thick line) – edge of C
(thin line) – edge of $G_{Y,d}^+$
(dashed line) – edge of $G_{Y,d}^-$

The arrow marks the direction in which the simplex algorithm passes the cycle. Short dotted lines indicate that $G_{Y,d}$ generally contains additional edges.

Fig. 4.2.2.1
Representation of the star-shaped graph S_C in Theorem 4.2.2.1

The tableaux $T_\nu, T_{\nu+1}, \bar{T}_\nu$, associated with $I_\nu, I_{\nu+1}, \bar{I}_\nu$, have the form in Tab. 4.2.2.1.
Since I_ν and $I_{\nu+1}$ are successive nodes of a simplex cycle, T_ν can be transformed into $T_{\nu+1}$ by a simplex step with the pivot $y'_{\sigma,1}$.

Tab. 4.2.2.1
Representation of the tableaux $T_\nu, T_{\nu+1}, \bar{T}_\nu$ in the proof of Theorem 4.2.2.1 (cf. Tab. 4.2.1.1)

T_ν	1	…		σ	$\sigma+1$	…	$\sigma+n$	$\sigma+n+1$
1	1				$y'_{1,1}$	…	$y'_{1,n}$	0
⋮		⋱			⋮			⋮
$\sigma-1$			1		⋮			⋮
σ				1	$\boxed{y'_{\sigma,1}}$	…	$y'_{\sigma,n}$	0
$\sigma+n+1$	0	…		0	$\boxed{d'_1}$	…	d'_n	1

$T_{\nu+1}$	1	…		σ	$\sigma+1$	…	$\sigma+n$	$\sigma+n+1$
1	1			$y''_{1,1}$	0	…	$y''_{1,n}$	0
⋮		⋱		⋮				⋮
$\sigma-1$			1	⋮				⋮
$\sigma+1$				$\frac{1}{y'_{\sigma,1}}$	1		$y''_{\sigma,n}$	0
$\sigma+n+1$	0	…	0	$\boxed{-\frac{d'_1}{y'_{\sigma,1}}}$	0	…	d''_n	1

$\bar{T}_\nu$	1	…		σ	$\sigma+1$	…	$\sigma+n$	$\sigma+n+1$
1	1				0		$y'''_{1,\sigma+n}$	$y'''_{1,\sigma+n+1}$
⋮		⋱			⋮			⋮
$\sigma-1$			1		⋮			⋮
σ				1	0	…	$y'''_{\sigma,\sigma+n}$	$-\frac{y'_{\sigma,1}}{d'_1}$
$\sigma+1$	0	…		0	1	…	d'''_n	$\frac{1}{d'_1}$

Thus $y'_{\sigma,1} > 0$ and $d'_1 < 0$. Consequently T_ν can be transformed into $\bar{T}_\nu$ by one pivot step with the pivot $d'_1 < 0$, i.e. (4.2.2.2) holds.

Analogously $T_{\nu+1}$ can be transformed into $\bar{T}_\nu$ by one pivot step with the pivot $-\frac{d'_1}{y'_{\sigma,1}} > 0$, i.e. (4.2.2.3) holds. Since

$\nu \in \{1,\dots,k\}$ may be chosen arbitrarily, condition (4.2.2.1) holds for the above defined nodes $\bar{I}_\nu$ $(\nu = 1,\dots,k)$.

b) We assume that there exist nodes $\bar{I}_\nu$ $(\nu = 1,\dots,k)$ satisfying (4.2.2.1).[120] We will now show that the nodes $\bar{I}_\nu$ are uniquely determined: Let $\nu \in \{1,\dots,k\}$ be a fixed index. Without loss of generality we may assume as in part a):

$$I_\nu = \{1,\dots,\sigma, n+\sigma+1\},$$

$$I_{\nu+1} = \{1,\dots,\sigma-1,\sigma+1,n+\sigma+1\}.$$

Since $\bar{I}_\nu$ is assumed to be not a node of G_Y^+ (cf. Remark 4.2.1.2),

$$\bar{I}_\nu \subset \{1,\dots,n+\sigma\} \tag{4.2.2.4}$$

holds. Moreover (4.2.2.1) implies

$$|\bar{I}_\nu \cap I_\nu| = \sigma \tag{4.2.2.5}$$

and

$$|\bar{I}_\nu \cap I_{\nu-1}| = \sigma \tag{4.2.2.6}$$

From (4.2.2.4) - (4.2.2.6) follows $\bar{I}_\nu = \{1,\dots,\sigma+1\}$ (cf. part a)).

Let the tableaux $T_\nu, T_{\nu+1}, \bar{T}_\nu$ be defined as in a) (cf. Tab. 4.2.2.1). Since the conditions (4.2.2.1) are assumed to be satisfied, $\bar{T}_\nu$ and T_ν can be transformed each into the other by a negative pivot step, while $\bar{T}_\nu$ and $T_{\nu+1}$ can be transformed each into the other by a positive pivot step. Thus $d_1' < 0$ and $-\frac{d_1'}{y_{\sigma,1}'} > 0$ hold, i.e. $d_1' < 0$ and $y_{\sigma,1}' > 0$. Hence we can transform T_ν into $T_{\nu+1}$ by a simplex step with the pivot $y_{\sigma,1}'$. Since $\nu \in \{1,\dots,k\}$ was chosen arbitrarily, C is a simplex cycle. •

Example 4.2.2.2:

Consider the reduced problem (4.1.6). In Fig. 4.2.2.2 the simplex cycle $C = (\{1,2\},\{2,3\},\{3,4\},\{4,5\}, \{5,6\},\{1,6\},\{1,2\})$ of (4.1.6) is

120 For the moment the nodes $\bar{I}_\nu$ defined here have nothing to do with the nodes $\bar{I}_\nu$ in part a) of the proof, but the following representations show that identical notations are justified.

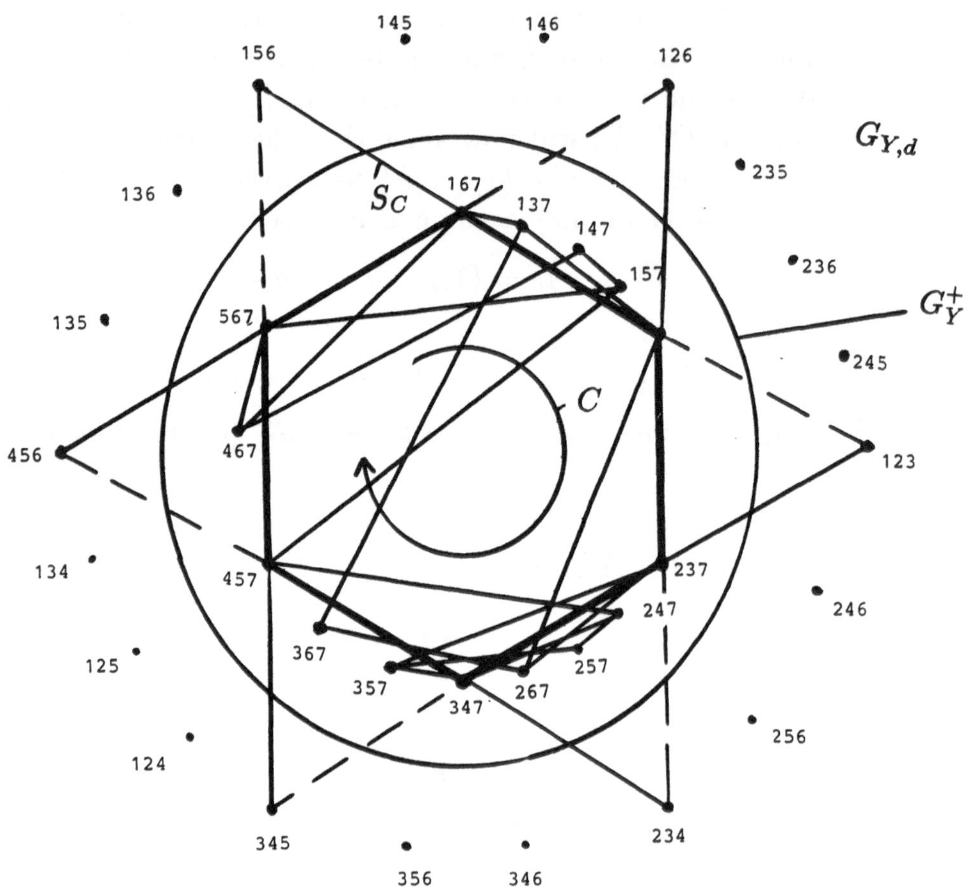

Legend:
$i\ j\ k$ – node $\{i,j,k\}$ of $G_{Y,d}$
– edge of C
– edge of $G_{Y,d}^+$
– edge of $G_{Y,d}^-$
– passing direction of the simplex algorithm

We identify the node $\{i,j,7\}$ of $G_{Y,d}$ with the node $\{i,j\}$ of G_Y^+ ($\{i,j\} \subset \{1,\ldots,6\}$, cf. Fig. 4.2.1.1). Because of the complexity of the LP-degeneracy graph, an edge of $G_{Y,d}$ is drawn only if it belongs to S_C or G_Y^+.

Fig. 4.2.2.2
Representation of the star-shaped graph $S_C \subset G_{Y,d}$ belonging to the simplex cycle C in Example 4.2.2.2

enlarged to the star-shaped graph S_C contained in $G_{Y,d}$ (cf. Example 4.1.3, Fig. 4.1.1).

Theorem 4.2.2.1 is the theoretical foundation for an algorithm detecting simplex cycles of a reduced problem (4.1.4), when the associated LP-degeneracy graph $G_{Y,d}$ is known. But for this purpose each cycle C of G_Y^+ has to be checked for "enlargability" to a star-shaped graph S_C. This involves considerable difficulties, since $G_{Y,d}$ in practice is very complex. Additional studies are necessary in order to prove whether an efficient algorithm can be developed on this basis.

4.3 GEOMETRICALLY MOTIVATED APPROACHES

In this section we characterize the simplex cycles of a reduced linear optimization problem by means of certain "geometrical" properties of the column vectors or of the vector of objective function coefficients in the initial tableau (cf. Theorems 4.3.2.2 and 4.3.4.4.).

4.3.1 FUNDAMENTALS

Let the reduced problem (4.1.4) and a basis B of (4.1.4) (regular $\sigma \times \sigma$-submatrix of $\bar{Y} \in \mathbb{R}^{\sigma \times \sigma + n}$) be given. For $j = 1, \ldots, n+\sigma$ and $i = 1, \ldots, \sigma$ we define the vectors $\tilde{y}^j, \beta_i, \tilde{d} \in \mathbb{R}^{\sigma+1}$ by

$$\left.\begin{array}{rcl} \tilde{y}^j & = & \binom{y^j}{d_j} \\ \beta_i & = & (\beta_i^{'}, 0), \\ \tilde{d} & = & (d_B^T B^{-1}, -1); \end{array}\right\} \tag{4.3.1.1}$$

with the notations
y^j – j-th column vector of $\bar{Y}$
d_j – j-th objective function coefficient of $\bar{d}$
$\beta_i^{'}$ – i-th row vector of B^{-1}
d_B^T – subvector of $\bar{d}$, belonging to basic indices
(d_B^T is a row vector).

The above notations make a "simple" representation of the simplex tableau possible.

Lemma 4.3.1.1:

The simplex tableau T associated with the basis B of (4.1.4) has the form

$$T = \begin{pmatrix} \beta_1 \tilde{y}^1 & ,\ldots, & \beta_1 \tilde{y}^{n+\sigma} \\ \vdots & & \vdots \\ \beta_\sigma \tilde{y}^1 & ,\ldots, & \beta_\sigma \tilde{y}^{n+\sigma} \\ \tilde{dy}^1 & ,\ldots, & \tilde{dy}^{n+\sigma} \end{pmatrix}$$

Proof: As is well known, T has the form

$$T = \left(\frac{B^{-1}\bar{Y}}{d_B^T B^{-1}\bar{Y} - \bar{d}} \right)$$

(cf. e.g. Gal (1983a:27)). Using the notations in (4.3.1.1) yields the assertion. •

For the rest of this publication we take the "usual" pivoting rules as a basis:[121]

- Any column with a *minimal (negative)* relative cost coefficient may be chosen as the pivot column.
- Any row with a *positive* pivot element may be taken as the pivot row.

Definition 4.3.1.2:

Let B and B^* denote bases of (4.1.4) with $B \overset{+}{\longleftrightarrow} B^*$. Moreover let j^+ denote the index of the entering basic variable when passing from B to B^*. The basis B^* is called a *successor* of B, if B can be transformed into B^* by one simplex step, i.e. if the following conditions are satisfied for the tableau T associated with B (cf. Lemma 4.3.1.1):

a) $\tilde{dy}^{j^+} < 0$

b) $\tilde{dy}^{j^+} \leq \tilde{dy}^j \quad \forall j = 1, \ldots, n+\sigma$

---

[121] Cf. the introductory representations in Section 4.2.2.

The following concepts are necessary for the characterization of simplex cycles in Section 4.3.2 (cf. Theorem 4.3.2.2):

Definition 4.3.1.3:

Let the reduced problem (4.1.4) and an (arbitrary) cycle $C = (I_1, \ldots, I_k, I_1)$ of G_Y^+ be given. For $\nu = 1, \ldots, k$ let $B_\nu(T_\nu)$ denote the basis (tableau) associated with I_ν. Moreover let $\bar{y}^j, \beta_{i,\nu}, \bar{d}_\nu$ denote the vectors used to represent T_ν (cf. (4.3.1.1), Lemma 4.3.1.1; $i = 1, \ldots, \sigma$; $j = 1, \ldots, n+\sigma$; $\nu = 1, \ldots, k$), and let j_ν^+ be the index of the entering basic variable when passing from B_ν to $B_{\nu+1}(B_{k+1} := B_1)$. Finally let H_ν and E_ν denote the closed halfspace

$$H_\nu = \{y \in \mathbb{R}^{\sigma+1} | \tilde{d}_\nu y \geq \tilde{d}_\nu \tilde{y}^{j_\nu^+}\} \tag{4.3.1.2}$$

of $\mathbb{R}^{\sigma+1}$ and the corresponding constraint-hyperplane

$$E_\nu = \{y \in \mathbb{R}^{\sigma+1} | \tilde{d}_\nu y = \tilde{d}_\nu \tilde{y}^{j_\nu^+}\}, \tag{4.3.1.3}$$

respectively.

The convex polyhedral set

$$P_C = \cap_{\nu=1}^k H_\nu \subset \mathbb{R}^{\sigma+1}$$

is called the *point set induced by C*.

Example 4.3.1.4:[122]

Let the reduced problem (4.1.4) have the special form

$$\left.\begin{array}{lllllll} \max z = & -5x_1 & -6x_2 & +x_3 & +2x_4 & & \\ \text{s.t.} & & & & & & \\ & 2x_1 & +3x_2 & +5x_3 & +x_4 & +x_5 & = 0 \\ & & & & & x_1, \ldots, x_5 & \geq 0 \end{array}\right\} \tag{4.3.1.4}$$

Moreover let the cycle $C = (\{1\}, \ldots, \{5\}, \{1\})$ of the positive degeneracy graph G_Y^+ be given ($Y = (2,3,5,1) \in \mathbb{R}^{1\times 4}$). This case yields

---

[122] Cf. Example 4.3.2.3.

$j_1^+ = 2, j_2^+ = 3, j_3^+ = 4, j_4^+ = 5, j_5^+ = 1,$
$\tilde{y}^{j_1^+} = \binom{3}{-6}, \tilde{y}^{j_2^+} = \binom{5}{1}, \tilde{y}^{j_3^+} = \binom{1}{2}, \tilde{y}^{j_4^+} = \binom{1}{0}, \tilde{y}^{j_5^+} = \binom{2}{-5},$
$\tilde{d}_1 = (-\frac{5}{2}, -1), \tilde{d}_2 = (-2, -1), \tilde{d}_3 = (\frac{1}{5}, -1),$
$\tilde{d}_4 = (2, -1), \tilde{d}_5 = (0, -1),$
$\tilde{d}_1 \tilde{y}^{j_1^+} = -\frac{3}{2}, \tilde{d}_2 \tilde{y}^{j_2^+} = -11, \tilde{d}_3 \tilde{y}^{j_3^+} = -\frac{9}{5},$
$\tilde{d}_4 \tilde{y}^{j_4^+} = 2, \tilde{d}_5 \tilde{y}^{j_5^+} = 5.$

In the above example H_ν and E_ν are the solution halfplane or its boundary line, respectively, associated with the ν-th inequality of the following system ($\nu = 1, \ldots, 5$):

$$\left.\begin{array}{llcrl} -\frac{5}{2}y_1 & -y_2 & \geq & -\frac{3}{2} & (1) \\ -2y_1 & -y_2 & \geq & -11 & (2) \\ \frac{1}{5}y_1 & -y_2 & \geq & -\frac{9}{5} & (3) \\ 2y_1 & -y_2 & \geq & 2 & (4) \\ & -y_2 & \geq & 5 & (5) \end{array}\right\} \qquad (4.3.1.5)$$

The point set P_C is the solution set of (4.3.1.5); it is illustrated in Fig. 4.3.1.1.

Example 4.3.1.5:[123]

Consider the reduced problem (4.1.6) with the cycle $C = (\{1,2\}, \{2,3\}, \{3,4\}, \{4,5\}, \{5,6\}, \{1,6\}, \{1,2\})$ of G_Y^+ (cf. Example 4.1.3). Analogously to the preceding example we obtain:

$y^{j_1^+} = 3, \qquad y^{j_2^+} = 4, \qquad y^{j_3^+} = 5, \qquad y^{j_4^+} = 6,$
$y^{j_5^+} = 1, \qquad y^{j_6^+} = 2,$

$$\tilde{y}^{j_1^+} = \begin{pmatrix} -\frac{1}{25} \\ -\frac{1}{50} \\ \frac{1}{50} \end{pmatrix}, \quad \tilde{y}^{j_2^+} = \begin{pmatrix} 9 \\ 3 \\ -6 \end{pmatrix}, \quad \tilde{y}^{j_3^+} = \begin{pmatrix} 1 \\ 0 \\ 0 \end{pmatrix}, \quad \tilde{y}^{j_4^+} = \begin{pmatrix} 0 \\ 1 \\ 0 \end{pmatrix},$$
$$\tilde{y}^{j_5^+} = \begin{pmatrix} \frac{1}{4} \\ \frac{1}{2} \\ \frac{3}{4} \end{pmatrix}, \quad \tilde{y}^{j_6^+} = \begin{pmatrix} -60 \\ -90 \\ -150 \end{pmatrix}.$$

Moreover it holds[124]

[123] Cf. Example 4.3.2.4.

[124] The first two components of the $\vec{d}_\nu$ are the relative cost coefficients in columns 5 and 6 of the respective tableaux in Tab. 4.1.2.

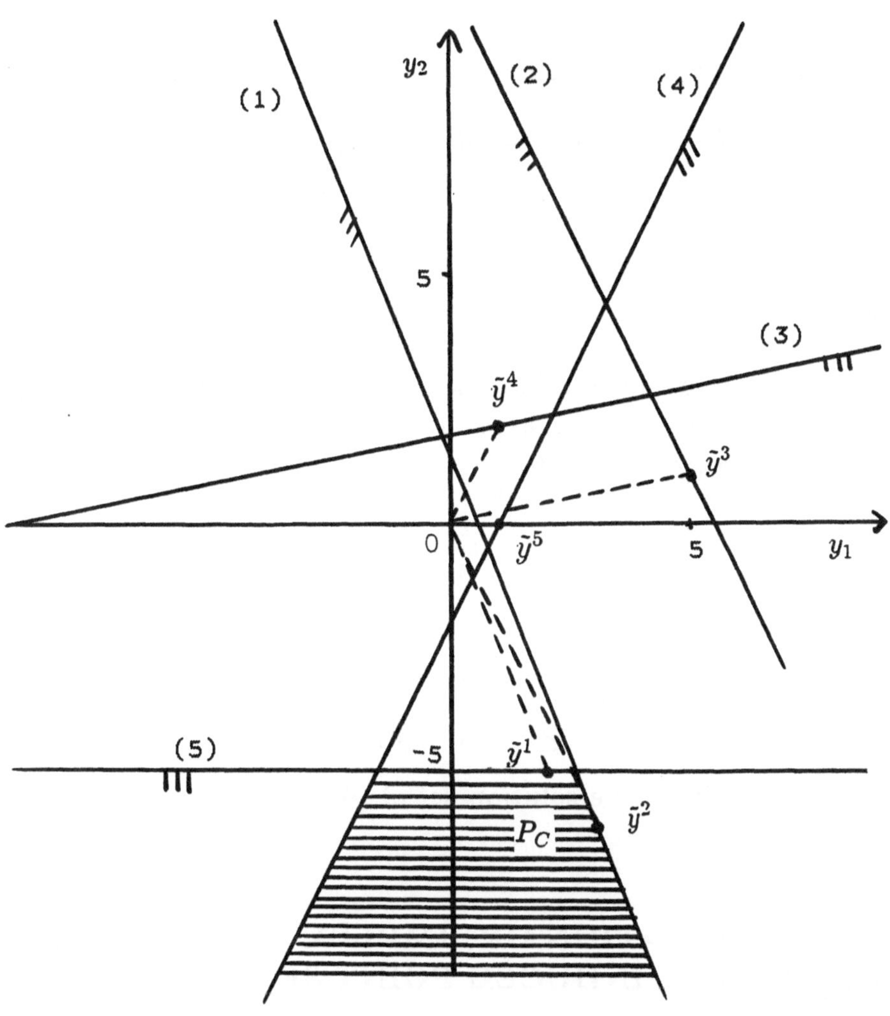

Legend:
(ν) – ν-th boundary line (E_ν)
The hatching at E_ν indicates the position of H_ν.

Fig. 4.3.1.1
Representation of the induced point set P_C and the vectors $\tilde{y}^\nu$ in Example 4.3.1.4

$$\tilde{d}_1 = ((\frac{3}{4}, -150) \cdot \begin{pmatrix} \frac{1}{4} & -60 \\ \frac{1}{2} & -90 \end{pmatrix}^{-1}, \quad -1) = (1, 1, -1),$$

$\tilde{d}_2 = (-2, 3, -1), \tilde{d}_3 = (-1, 1, -1), \tilde{d}_4 = (0, -2, -1), \tilde{d}_5 = (0, 0, -1), \tilde{d}_6 = (3, 0, -1)$.

This yields

$\tilde{d}_1 \tilde{y}^{j_1^+} = -\frac{4}{50}, \tilde{d}_2 \tilde{y}^{j_2^+} = -3, \tilde{d}_3 \tilde{y}^{j_3^+} = -1, \tilde{d}_4 \tilde{y}^{j_4^+} = -2, \tilde{d}_5 \tilde{y}^{j_5^+} = -\frac{3}{4}, \tilde{d}_6 \tilde{y}^{j_6^+} = -30.$

Hence H_ν and E_ν are the solution halfspace or the corresponding constraint-hyperplane, respectively, associated with the ν-th inequality of the following system ($\nu = 1, \ldots, 6$):

$$\left.\begin{array}{rrrcrl} y_1 & +y_2 & -y_3 & \geq & -\frac{4}{50} & (1) \\ -2y_1 & +3y_2 & -y_3 & \geq & -3 & (2) \\ -y_1 & +y_2 & -y_3 & \geq & -1 & (3) \\ & -2y_2 & -y_3 & \geq & -2 & (4) \\ & & -y_3 & \geq & -\frac{3}{4} & (5) \\ 3y_1 & & -y_3 & \geq & -30 & (6) \end{array}\right\} \qquad (4.3.1.6)$$

The induced point set P_C is the solution set of (4.3.1.6).

4.3.2 CHARACTERIZATION OF SIMPLEX CYCLES BY MEANS OF THE INDUCED POINT SET

The characterization below is based on

Lemma 4.3.2.1:

Let the reduced problem (4.1.4) and a cycle $C = (I_1, \ldots, I_k, I_1)$ of G_Y^+ be given. The cycle C is a simplex cycle if and only if the following conditions are satisfied:[125]

a) $\tilde{d}_\nu \tilde{y}^{j_\nu^+} < 0 \;\; \forall \nu = 1, \ldots, k$

b) $\tilde{d}_\nu \tilde{y}^{j_\nu^+} \leq \tilde{d}_\nu \tilde{y}^j \;\; \forall \nu = 1, \ldots, k \quad , \quad \forall j = 1, \ldots, n + \sigma.$

---

[125] Cf. the notations in (4.3.1.1), Lemma 4.3.1.1 and Def. 4.3.1.3.

Proof: The above conditions mean that $B_{\nu+1}$ is a successor of B_ν for $\nu = 1,\ldots,k$ (cf. Def. 4.3.1.2; B_ν denotes the basis associated with I_ν). From this follows the assertion. •

The concept of the induced point set makes a geometrical interpretation of the above lemma possible:

Theorem 4.3.2.2:

Given (4.1.4) and the cycle C of G_Y^+ in Lemma 4.3.2.1. The cycle C is a simplex cycle of (4.1.4) if and only if the vectors $0, \tilde{y}^j \in \mathbb{R}^{\sigma+1}$ (cf. (4.3.1.1)) satisfy the following conditions[126]:

a) $0 \in \text{ int } P_C$

b) $\tilde{y}^j \in P_C \;\; \forall j = 1,\ldots,n+\sigma$.

Proof: We will show that the respective conditions in Lemma 4.3.2.1 and Theorem 4.3.2.2 are equivalent:

The condition a) in Lemma 4.3.2.1 means that the vector $0 \in \mathbb{R}^{\sigma+1}$ belongs to the interior of all halfspaces H_ν ($\nu = 1,\ldots,k$;cf. (4.3.1.2)), i.e. it is equivalent to the condition a) of the assertion.

The condition b) in Lemma 4.3.2.1 means that

$$\tilde{y}^j \in H_\nu$$

for $\nu = 1,\ldots,k$ and $j = 1,\ldots,n+\sigma$, i.e. all vectors $\tilde{y}^j$ ($j = 1,\ldots,n+\sigma$) are elements of P_C. Thus the respective conditions b) in Lemma 4.3.2.1 and Theorem 4.3.2.2 are also equivalent. •

In order to decide whether a cycle C of a positive degeneracy graph G_Y^+ is a simplex cycle, we must prove whether the vectors 0, $\tilde{y}^j$ are elements of P_C ($0 \in \text{ int } P_C$). It is important to note that P_C *depends on the vectors* $\tilde{y}^j$.

Let us illustrate Theorem 4.3.2.2 by means of two examples:

Example 4.3.2.3:

Consider the reduced problem (4.3.1.4) with the cycle C in Example 4.3.1.4. Now Fig. 4.3.1.1 shows that the vector $0 \in \mathbb{R}^2$ as well as

---

[126] Cf. Appendix, Def. A.10.

the vectors $\tilde{y}^3, \tilde{y}^4, \tilde{y}^5$ lie outside of the point set P_C. Theorem 4.3.2.2 implies that C is not a simplex cycle.

Example 4.3.2.4:

Given the reduced problem (4.1.6) with the cycle C in Example 4.3.1.5. The vector $0 \in I\!R^3$ belongs to the interior of P_C, since all components of the right-hand side in (4.3.1.6) are negative. Inserting the vectors $\tilde{y}^j$ in (4.3.1.6) shows that all $\tilde{y}^j$ are elements of P_C. As Theorem 4.3.2.2 implies, C is not a simplex cycle.

The use of Theorem 4.3.2.2 depends on whether a relative "simple" representation of P_C is possible. That is the reason we investigate P_C in the following section.

4.3.3 PROPERTIES OF THE INDUCED POINT SET

From Fig. 4.3.1.1 we draw the following properties of the induced point set P_C in Example 4.3.1.4:

a) The ν-th boundary line E_ν runs parallel with the subspace $U(\tilde{y}^\nu)$ induced by the vector $\tilde{y}^\nu$ and contains the point $\tilde{y}^{j_\nu^+}$, i.e. E_ν is an affine subspace of the form[127]

$$E_\nu = U(\tilde{y}^{j_\nu^+}; \tilde{y}^\nu)(\nu = 1, \ldots, 5).$$

b) The halfplane H_ν associated with E_ν lies below E_ν $(\nu = 1, \ldots, 5)$.
c) The point set P_C is unbounded.

It is possible to transfer the above assertions on P_C to the general case:

Theorem 4.3.3.1:

Let the reduced problem (4.1.4) and a cycle $C = (I_1, \ldots, I_k, I_1)$ of G_Y^+ be given. Let the index set I_ν of C have the form $I_\nu = \{j_1^{(\nu)}, \ldots, j_\sigma^{(\nu)}\}$ $(\nu = 1, \ldots, k)$.

---

127 Cf. Appendix, Def. A.14.

Then the induced point set P_C has the following properties:[128]

a) The ν-th constraint-hyperplane E_ν runs parallel to $U(\tilde{y}^{j_1^{(\nu)}},\ldots,\tilde{y}^{j_\sigma^{(\nu)}})$ and contains $\tilde{y}^{j_\nu^+}$,i.e.

$$E_\nu = U(\tilde{y}^{j_\nu^+};\tilde{y}^{j_1^{(\nu)}},\ldots,\tilde{y}^{j_\sigma^{(\nu)}})\ (\nu = 1,\ldots,k). \tag{4.3.3.1}$$

b) The halfspace H_ν associated with E_ν lies "below"[129] E_ν, i.e. that

$$-e_{\sigma+1} \notin U(\tilde{y}^{j_1^{(\nu)}},\ldots,y^{j_\sigma^{(\nu)}}), \tag{4.3.3.2}$$

and for a sufficiently large $\alpha_\nu \in I\!R, \alpha_\nu > 0$ holds

$$\{-\alpha e_{\sigma+1} | \alpha \geq \alpha_\nu\} \subset H_\nu \quad (\nu = 1,\ldots,k). \tag{4.3.3.3}$$

c) The point set P_C is always unbounded.

Proof:

a) From (4.3.1.2) follows $E_\nu = \{y \in I\!R^{\sigma+1} | \tilde{d}_\nu y = \tilde{d}_\nu \tilde{y}^{j_\nu^+}\}$, i.e. $E_\nu = \tilde{y}^{j_\nu^+} + \{y \in I\!R^{\sigma+1} | \tilde{d}_\nu y = 0\}$. In order to prove part a) of the assertion we show that

$$\{y \in I\!R^{\sigma+1} | \tilde{d}_\nu y = 0\} = U(\tilde{y}^{j_1^{(\nu)}},\ldots,\tilde{y}^{j_\sigma^{(\nu)}})(\nu = 1,\ldots,k) \tag{4.3.3.4}$$

holds.

Now for $\nu = 1,\ldots,k$ and $\mu = 1,\ldots,\sigma$ holds[130] (cf. (4.3.1.1)):

$$\begin{aligned}
\tilde{d}_\nu \tilde{y}^{j_\mu^{(\nu)}} &= (d_{B_\nu}^T B_\nu^{-1}, -1)\binom{y^{j_\mu^{(\nu)}}}{d_{j_\mu^{(\nu)}}} \\
&= d_{B_\nu}^T B_\nu^{-1} y^{j_\mu^{(\nu)}} - d_{j_\mu^{(\nu)}} \\
&= (d_{j_1^{(\nu)}},\ldots,d_{j_\sigma^{(\nu)}})(y^{j_1^{(\nu)}}|\ldots|y^{j_\sigma^{(\nu)}})^{-1} y^{j_\mu^{(\nu)}} - d_{j_\mu^{(\nu)}} \\
&= (d_{j_1^{(\nu)}},\ldots,d_{j_\sigma^{(\nu)}}) e_\mu - d_{j_\mu^{(\nu)}} \\
&= d_{j_\mu^{(\nu)}} - d_{j_\mu^{(\nu)}} = 0.
\end{aligned}$$

---

[128] Cf. the notations in Def. 4.3.1.3.

[129] "Below" E_ν means that H_ν lies at that side of E_ν, in which the vector $-e_{\sigma+1}$ directs.

[130] B_ν denotes the basis associated with I_ν.

Thus the σ-dimensional subspace $\{y \in \mathbb{R}^{\sigma+1} | \tilde{d}_\nu y = 0\}$ contains the linearly independent vectors $\tilde{y}^{j_\mu^{(\nu)}}$ ($\mu = 1, \ldots, \sigma$), i.e. (4.3.3.4) holds.

b) Let $\nu \in \{1, \ldots, k\}$ be fixed. Because of rank $(-e_{\sigma+1} | \tilde{y}^{j_1^{(\nu)}} | \ldots | \tilde{y}^{j_\sigma^{(\nu)}}) =$

$$\operatorname{rank} \left(\begin{array}{c|c} 0 & \\ \vdots & B_\nu \\ 0 & \\ \hline -1 & d_{j_1^{(\nu)}}, \ldots, d_{j_\sigma^{(\nu)}} \end{array} \right) = \sigma + 1$$

the vector $-e_{\sigma+1}$ is not a linear combination of the vectors $\tilde{y}^{j_1^{(\nu)}}, \ldots, \tilde{y}^{j_\sigma^{(\nu)}}$, i.e. (4.3.3.2) holds.
Let be $\alpha_\nu := \tilde{d}_\nu \tilde{y}^{j_\nu^+}$. Then for $\alpha \geq \alpha_\nu$ holds (cf. Def. 4.3.1.3 and (4.3.1.1)):

$$\tilde{d}_\nu(-\alpha e_{\sigma+1}) = \alpha \geq \tilde{d}_\nu \tilde{y}^{j_\nu^+},$$

i.e. $-\alpha e_{\sigma+1} \in H_\nu$ (cf. (4.3.1.2)). Thus (4.3.3.3) holds.

c) Part b) of the assertion implies

$$\{-\alpha e_{\sigma+1} | \alpha \geq \alpha_0\} \subset \cap_{\nu=1}^{k} H_\nu = P_C$$

for $\alpha_0 := \max\{\alpha_\nu | \nu = 1, \ldots, k\}$.
Hence P_C is unbounded. •

Theorem 4.3.3.1 enables us to represent the induced point set P_C or the associated halfspaces H_ν and constraint-hyperplanes E_ν by the vectors $\tilde{y}^j$ (cf. (4.3.1.1)).

Example 4.3.3.2:

For the induced point set P_C in Example 4.3.1.5 (cf. (4.3.1.6)) the constraint-hyperplanes are as follows:

$$\begin{array}{ll} E_1 = (\tilde{y}^3; \tilde{y}^1, \tilde{y}^2) & E_2 = (\tilde{y}^4; \tilde{y}^2, \tilde{y}^3) \\ E_3 = (\tilde{y}^5; \tilde{y}^3, \tilde{y}^4) & E_4 = (\tilde{y}^6; \tilde{y}^4, \tilde{y}^5) \\ E_5 = (\tilde{y}^1; \tilde{y}^5, \tilde{y}^6) & E_6 = (\tilde{y}^2; \tilde{y}^1, \tilde{y}^6) \end{array}$$

This means that E_1 is the plane parallel to the subspace generated by $\tilde{y}^1 = (\frac{1}{4}, \frac{1}{2}, \frac{3}{4})^T$ and $\tilde{y}^2 = (-60, -90, -150)^T$, which contains $\tilde{y}^3 = (-\frac{1}{25}, -\frac{1}{50}, \frac{1}{50})^T$, etc. The halfspace H_ν lies below $E_\nu (\nu = 1, \ldots, 6)$.

Using Theorem 4.3.3.1 we can state the following corollary to Theorem 4.3.2.2 which characterizes simplex cycles by certain relations between the vectors $\tilde{y}^j$ (cf. (4.3.1.1)).

Corollary 4.3.3.3:

On the basis of the assumptions and notations in Theorem 4.3.3.1 holds: The cycle C is a simplex cycle if and only if the vectors $\tilde{y}^j$ satisfy the following conditions:

a) There exist real numbers $\alpha_1^{(\nu)}, \ldots, \alpha_\sigma^{(\nu)}, \beta^{(\nu)}; \beta^{(\nu)} > 0$ with
$0 = \tilde{y}^{j_\nu^+} + \alpha_1^{(\nu)} \tilde{y}^{j_1^{(\nu)}} + \ldots + \alpha_\sigma^{(\nu)} \tilde{y}^{j_\sigma^{(\nu)}} - \beta^{(\nu)} e_{\sigma+1}$
$(\nu = 1, \ldots, k)$

b) There exist real numbers $\alpha_1^{(\nu,j)}, \ldots, \alpha_\sigma^{(\nu,j)}, \beta^{(\nu,j)}; \beta^{(\nu,j)} \geq 0$ with
$\tilde{y}^j = \tilde{y}^{j_\nu^+} + \alpha_1^{(\nu,j)} \tilde{y}^{j_1^{(\nu)}} + \ldots + \alpha_\sigma^{(\nu,j)} \tilde{y}^{j_\sigma^{(\nu)}} - \beta^{(\nu,j)} e_{\sigma+1}$
$(\nu = 1, \ldots, k; j = 1, \ldots, n + \sigma)$

Proof: The condition a) means that the vector $0 \in \mathbb{R}^{\sigma+1}$ lies "below" all constraint-hyperplanes of P_C, i.e. it is equivalent to condition a) in Theorem 4.3.2.2. The condition b) means that all vectors $\tilde{y}^j$ lie "below" (or on) all constraint-hyperplanes of P_C, i.e. it is equivalent to condition b) in Theorem 4.3.2.2. •

According to the above corollary a cycle C of G_Y^+ is a simplex cycle if and only if a system of $(\sigma+1) \cdot (k + k \cdot (n + \sigma)) = (n + \sigma + 1)(\sigma + 1) \cdot k$ nonlinear equations is satisfied.

We could use this result in particular for the construction of cycling examples. However, with regard to the procedures in Chapter 5, which construct cycling examples based on other considerations, we will not pursue this idea.

4.3.4 CHARACTERIZATION OF SIMPLEX CYCLES BY MEANS OF THE INDUCED CONE

The starting point of the following is question (4.1.7) in a slightly modified form. The *system of constraints* in (4.1.4) as well as a *cycle C* of

G_Y^+ are assumed to be *fixed*. We now ask, which necessary and sufficient conditions the vector d must satisfy such that C is a simplex cycle.

Definition 4.3.4.1:

Let the system of constraints in (4.1.4) and an (arbitrary) cycle $C = (I_1, \ldots, I_k, I_1)$ of G_Y^+ be given. The set of all vectors $d \in \mathbb{R}^n$ satisfying the homogeneous linear system of inequalities

$$\left.\begin{array}{rr} (B_\nu^{-1} y^{j_\nu^+})^T d_{B_\nu} - d_{j_\nu^+} < 0 & \forall \nu = 1, \ldots, k \\ (B_\nu^{-1} y^{j_\nu^+})^T d_{B_\nu} - d_{j_\nu^+} \leq (B_\nu^{-1} y^j)^T d_{B_\nu} - d_j & \forall \nu = 1, \ldots, k \\ & \forall j = 1, \ldots, n+\sigma \end{array}\right\} \tag{4.3.4.1}$$

is denoted by K_C $(d_{n+1} = \ldots = d_{n+\sigma} = 0)$.

Since (4.3.4.1) is homogeneous, the set K_C is a convex cone, called the *cone induced by* C (cf. Lemma A.8).

Example 4.3.4.2:

Let the system of constraints in (4.1.6) and the cycle C in Example 4.3.1.5 be given. After certain redundant inequalities have been omitted, the system (4.3.4.1) can be transformed into (4.3.4.2).[131]
The corresponding "normal form" is (4.3.4.3).
Hence the cone K_C is the solution set of (4.3.4.3). From $(d_1, d_2, d_3, d_4) = (\frac{3}{4}, -150, \frac{1}{50}, -6) \in K_C$ follows that $K_C \neq \emptyset$.

---

131 Note that $d_5 = d_6 = 0$. The vector $B_\nu^{-1} y^j$ is the j-th column of the tableau of (4.1.6) associated with B_ν (cf. the subtableaux of Tab. 4.1.2 in bold type).

$$\left.\begin{array}{llll}
\frac{8}{25}d_1 & +\frac{1}{500}d_2 & -d_3 & \left\{\begin{array}{lllll} < & 0 & & & (\nu = 1) \\ \leq & -84d_1 & -\frac{1}{2}d_2 & -d_4 & (\nu = 1, \quad j = 4) \\ \leq & -12d_1 & -\frac{1}{15}d_2 & -d_5 & (\nu = 1, \quad j = 5) \\ \leq & 8d_1 & +\frac{1}{30}d_2 & -d_6 & (\nu = 1, \quad j = 6) \end{array}\right. \\
\frac{1}{40}d_2 & -\frac{525}{2}d_3 & -d_4 & \left\{\begin{array}{lllll} < & 0 & & & (\nu = 2) \\ \leq & -\frac{1}{160}d_2 & +\frac{25}{8}d_3 & -d_1 & (\nu = 2, \quad j = 1) \\ \leq & \frac{1}{120}d_2 & -\frac{75}{2}d_3 & -d_5 & (\nu = 2, \quad j = 5) \\ \leq & -\frac{1}{60}d_2 & +25d_3 & -d_6 & (\nu = 2, \quad j = 6) \end{array}\right. \\
50d_3 & +\frac{1}{3}d_4 & -d_5 & \left\{\begin{array}{lllll} < & 0 & & & (\nu = 3) \\ \leq & -\frac{125}{2}d_3 & -\frac{1}{4}d_4 & -d_1 & (\nu = 3, \quad j = 1) \\ \leq & 10500d_3 & +40d_4 & -d_2 & (\nu = 3, \quad j = 2) \\ \leq & -150d_3 & -\frac{2}{3}d_4 & -d_6 & (\nu = 3, \quad j = 6) \end{array}\right. \\
\frac{1}{3}d_4 & -3d_5 & -d_6 & \left\{\begin{array}{lllll} < & 0 & & & (\nu = 4) \\ \leq & \frac{1}{6}d_4 & -\frac{5}{4}d_5 & -d_1 & (\nu = 4, \quad j = 1) \\ \leq & -30d_4 & +210d_5 & -d_2 & (\nu = 4, \quad j = 2) \\ \leq & -\frac{1}{150}d_4 & +\frac{1}{50}d_5 & -d_3 & (\nu = 4, \quad j = 3) \end{array}\right. \\
\frac{1}{4}d_5 & +\frac{1}{2}d_6 & -d_1 & \left\{\begin{array}{lllll} < & 0 & & & (\nu = 5) \\ \leq & -60d_5 & -90d_6 & -d_2 & (\nu = 5, \quad j = 2) \\ \leq & -\frac{1}{25}d_5 & -\frac{1}{50}d_6 & -d_3 & (\nu = 5, \quad j = 3) \\ \leq & 9d_5 & +3d_6 & -d_4 & (\nu = 5, \quad j = 4) \end{array}\right. \\
-240d_1 & +30d_6 & -d_2 & \left\{\begin{array}{lllll} < & 0 & & & (\nu = 5) \\ \leq & -\frac{4}{25}d_1 & +\frac{3}{50}d_6 & -d_3 & (\nu = 6, \quad j = 3) \\ \leq & 36d_1 & -15d_6 & -d_4 & (\nu = 5, \quad j = 4) \\ \leq & 4d_1 & -2d_6 & -d_5 & (\nu = 6, \quad j = 5) \end{array}\right.
\end{array}\right\} \tag{4.3.4.2}$$

$$
\left.
\begin{array}{rlrlrlrlrll}
\frac{8}{25} & d_1+ & \frac{1}{500} & d_2- & & d_3 & & & < 0 & (\nu = 1) & \\
(\frac{8}{25} + 84) & d_1+ & (\frac{1}{500} + \frac{1}{2}) & d_2- & & d_3+ & & d_4 & \leq 0 & (\nu = 1, & j = 4) \\
(\frac{8}{25} + 12) & d_1+ & (\frac{1}{500} + \frac{1}{15}) & d_2- & & d_3 & & & \leq 0 & (\nu = 1, & j = 5) \\
(\frac{8}{25} - 8) & d_1+ & (\frac{1}{500} - \frac{1}{30}) & d_2- & & d_3 & & & \leq 0 & (\nu = 1, & j = 6) \\
& & \frac{1}{40} & d_2- & \frac{525}{2} & d_3- & & d_4 & < 0 & (\nu = 2) & \\
& d_1+ & (\frac{1}{40} + \frac{1}{160}) & d_2- & (\frac{525}{2} + \frac{25}{8}) & d_3- & & d_4 & \leq 0 & (\nu = 2, & j = 1) \\
& & (\frac{1}{40} - \frac{1}{120}) & d_2+ & (\frac{75}{2} - \frac{525}{2}) & d_3- & & d_4 & \leq 0 & (\nu = 2, & j = 5) \\
& & (\frac{1}{40} + \frac{1}{60}) & d_2- & (\frac{525}{2} + 25) & d_3- & & d_4 & \leq 0 & (\nu = 2, & j = 6) \\
& & & & 50 & d_3+ & \frac{1}{3} & d_4 & < 0 & (\nu = 3) & \\
& d_1+ & & & (50 + \frac{125}{2}) & d_3+ & (\frac{1}{3} + \frac{1}{4}) & d_4 & \leq 0 & (\nu = 3, & j = 1) \\
& & & d_2+ & (50 - 10500) & d_3+ & (\frac{1}{3} - 40) & d_4 & \leq 0 & (\nu = 3, & j = 2) \\
& & & & (50 + 150) & d_3+ & (\frac{1}{3} + \frac{2}{3}) & d_4 & \leq 0 & (\nu = 3, & j = 6) \\
& & & & & & \frac{1}{3} & d_4 & < 0 & (\nu = 4) & \\
& d_1+ & & & & & (\frac{1}{3} - \frac{1}{6}) & d_4 & \leq 0 & (\nu = 4, & j = 1) \\
& & & d_2+ & & & (\frac{1}{3} + 30) & d_4 & \leq 0 & (\nu = 4, & j = 2) \\
& & & & & d_3+ & (\frac{1}{3} + \frac{1}{150}) & d_4 & \leq 0 & (\nu = 4, & j = 3) \\
& -d_1 & & & & & & & < 0 & (\nu = 5) & \\
& -d_1+ & & d_2 & & & & & \leq 0 & (\nu = 5, & j = 2) \\
& -d_1+ & & & & d_3 & & & \leq 0 & (\nu = 5, & j = 3) \\
& -d_1+ & & & & & & d_4 & \leq 0 & (\nu = 5, & j = 4) \\
-240 & d_1- & & d_2 & & & & & < 0 & (\nu = 6) & \\
(-240 + \frac{4}{25}) & d_1- & & d_2+ & & d_3 & & & \leq 0 & (\nu = 6, & j = 3) \\
(-240 - 36) & d_1- & & d_2+ & & & & d_4 & \leq 0 & (\nu = 6, & j = 4) \\
(-240 - 4) & d_1- & & d_2 & & & & & \leq 0 & (\nu = 6, & j = 5)
\end{array}
\right\}
\tag{4.3.4.3}
$$

The case $K_C = \emptyset$ occurs in

Example 4.3.4.3:

Let the system of constraints in (4.1.4) be of the form

$$\left.\begin{array}{llll} 2x_1 & & +x_3 & = 0 \\ & 2x_2 & \quad +x_4 & = 0 \\ & & x_1, \ldots, x_4 & \geq 0 \end{array}\right\} \tag{4.3.4.4}$$

Let $C = (\{1,2\},\{2,3\},\{3,4\},\{1,4\},\{1,2\})$ be the given cycle of the positive degeneracy graph G_Y^+. After certain redundant inequalities have been omitted, the system (4.3.4.1) has the form ($d_3 = d_4 = 0$):

$$\left.\begin{array}{ll} \frac{1}{2}d_1 \begin{cases} < 0 & (\nu = 1) \\ \leq \frac{1}{2}d_2 & (\nu = 1, j = 4) \end{cases} \\ \frac{1}{2}d_2 \begin{cases} < 0 & (\nu = 2) \\ \leq -d_1 & (\nu = 2, j = 1) \end{cases} \\ -d_1 \begin{cases} < 0 & (\nu = 3) \\ \leq -d_2 & (\nu = 3, j = 2) \end{cases} \\ -d_2 \begin{cases} < 0 & (\nu = 4) \\ \leq \frac{1}{2}d_1 & (\nu = 4, j = 3) \end{cases} \end{array}\right\} \tag{4.3.4.5}$$

By means of the inequalities "$\nu = 1$" and "$\nu = 3$" we see immediately that system (4.3.4.5) is inconsistent. From this follows $K_C = \emptyset$.

The concept of the induced cone leads to the following variant of Theorem 4.3.2.2:

Theorem 4.3.4.4:

Let the reduced problem (4.1.4) and a cycle C of G_Y^+ be given, where Y and C are fixed and d is variable. The cycle C is a simplex cycle of (4.1.4) if and only if the vector d of objective function coefficients satisfies the condition

$$d \in K_C \tag{4.3.4.6}$$

Proof: Using (4.3.1.1) and (4.3.1.2) we can transform the conditions in (4.3.4.1) as follows (cf. also Def. 4.3.1.3):

a)

$$\begin{array}{lll} (B_\nu^{-1}y^{j_\nu^+})^T \quad d_{B_\nu} - d_{j_\nu^+} & <0 & \forall \nu = 1,\ldots,k \quad \Leftrightarrow \\ \tilde{d}_\nu \tilde{y}^{j_\nu^+} & <0 & \forall \nu = 1,\ldots,k \quad \Leftrightarrow \end{array}$$

$$0 \in \cap_{\nu=1}^{k} H_\nu = P_C$$

b)

$$\begin{array}{lll} (B_\nu^{-1}y^{j_\nu^+})^T d_{B_\nu} - d_{j_\nu^+} & \leq (B_\nu^{-1}y^{j})^T d_{B_\nu} - d_j & \forall \nu = 1,\ldots,k \\ & & \forall j = 1,\ldots,n+\sigma \quad \Leftrightarrow \end{array}$$

$$\begin{array}{llll} \tilde{d}_\nu \tilde{y}^{j_\nu^+} & \leq \tilde{d}_\nu y^j & \forall \nu = 1,\ldots,k; & \forall j = 1,\ldots,n+\sigma \quad \Leftrightarrow \\ \tilde{y}^j & \in H_\nu & \forall \nu = 1,\ldots,k; & \forall j = 1,\ldots,n+\sigma \quad \Leftrightarrow \\ \tilde{y}^j & \in P_C & \forall j = 1,\ldots,n+\sigma & \end{array}$$

Hence the "$<$"- and "$\leq$"- conditions in (4.3.4.1) are equivalent to the conditions a) and b) in Theorem 4.3.2.2. •

Example 4.3.4.5:

Let the reduced problem

$$\left.\begin{array}{llllllll} \max z = & d_1x_1 & +d_2x_2 & +d_3x_3 & +d_4x_4 & & & \\ \text{s.t.} & & & & & & & \\ & \frac{1}{4}x_1 & -60x_2 & -\frac{1}{25}x_3 & +9x_4 & +x_5 & & = 0 \\ & \frac{1}{2}x_1 & -90x_2 & -\frac{1}{50}x_3 & +3x_4 & & +x_6 & = 0 \\ & & & & & & x_1,\ldots,x_6 & \geq 0 \end{array}\right\} \quad (4.3.4.7)$$

be given (cf. (4.1.6)). Let the vector of objective function coefficients be alternatively given as:

a) $d = d^1 := (\frac{3}{4}, -150, \frac{1}{50}, -6)^T$
b) $d = d^2 := (\frac{1}{8}, -\frac{2976}{100}, \frac{3}{100}, -\frac{13}{2})^T$
c) $d = d^3 := (2, -390, \frac{43}{50}, 26)^T$

In each case let

$$C = (\{1,2\}, \{2,3\}, \{3,4\}, \{4,5\}, \{5,6\}, \{1,6\}, \{1,2\}) \qquad (4.3.4.8)$$

be the given cycle of the positive degeneracy graph G_Y^+.

For the above values of the vector d we obtain the following (cf. Example 4.3.4.2):

a) It holds $d^1 \in K_C$, since d^1 satisfies the system (4.3.4.3). From Theorem 4.3.4.4 follows that C is a simplex cycle of (4.3.4.7) for $d = d^1$.
b) It holds $d^2 \in K_C$, i.e. C is a simplex cycle of (4.3.4.7) for $d = d^2$. The associated tableaux are listed in Tab. 4.3.4.1.
c) Since d^3 does not satisfy the inequality "$\nu = 3$" in (4.3.4.3), it holds $d^3 \notin K_C$. Thus C is *not* a simplex cycle of (4.3.4.7) for $d = d^3$.

The fact that the inequality "$\nu = 3$" is not satisfied means that the basis B_4 is not a successor of B_3.[132]. We can show this by means of the tableau associated with B_3 which has the form

	1	2	3	4	5	6	x_B
3	$-\frac{125}{2}$	10500	1	0	50	-150	0
4	$-\frac{1}{4}$	40	0	1	$\frac{1}{3}$	$-\frac{2}{3}$	0
Δz_j	$-\frac{249}{4}$	10460	0	0	$\frac{155}{3}$	$-\frac{439}{3}$	0

(4.3.4.9)

Since the columns with negative relative costs contain no positive elements, B_3 has no successor.

132 B_3 and B_4 denote the bases associated with $I_3 = \{3,4\}$ and $I_4 = \{4,5\}$, respectively (cf. (4.3.4.8) and Def. 4.3.1.3).

Tab. 4.3.4.1
Tableaux associated with the simplex cycle C in Example 4.3.4.5b)

	1	2	3	4	5	6
5	$\frac{1}{4}$	- 60	$-\frac{1}{25}$	9	1	0
6	$\frac{1}{2}$	-90	$-\frac{1}{50}$	3	0	1
Δz_j	$-\frac{1}{8}$	29,76	$-\frac{3}{100}$	$\frac{13}{2}$	0	0
1	1	-240	$-\frac{4}{25}$	36	4	0
6	0	30	$\frac{3}{50}$	-15	-2	1
Δz_j	0	-0,24	$-\frac{1}{20}$	11	$\frac{1}{2}$	0
1	1	0	$\frac{8}{25}$	-84	-12	8
2	0	1	$\frac{1}{500}$	$-\frac{1}{2}$	$-\frac{1}{15}$	$\frac{1}{30}$
Δz_j	0	0	-0,04952	10,88	0,484	0,008
3	$\frac{25}{8}$	0	1	$-\frac{525}{2}$	$-\frac{75}{2}$	25
2	$-\frac{1}{160}$	1	0	$\frac{1}{40}$	$\frac{1}{120}$	$-\frac{1}{60}$
Δz_j	0,15475	0	0	-2,119	-1,373	1,246
3	$-\frac{125}{2}$	10500	1	0	50	-150
4	$-\frac{1}{4}$	40	0	1	$\frac{1}{3}$	$-\frac{2}{3}$
Δz_j	-0,375	84,76	0	0	$-\frac{2}{3}$	$-\frac{1}{6}$
5	$-\frac{5}{4}$	210	$\frac{1}{50}$	0	1	-3
4	$\frac{1}{6}$	-30	$-\frac{1}{150}$	1	0	$\frac{1}{3}$
Δz_j	$-\frac{3,625}{3}$	224,76	$\frac{1}{75}$	0	0	$-\frac{13}{6}$

The next pivot step generates the initial tableau.

Legend: Tableau associated with basis B

	column indices
indices of basis B	$B^{-1}\bar{Y}$
Δz_j	$d_B^T B^{-1}\bar{Y} - \bar{d}$

(The right-hand side always equals zero and is therefore omitted).

4.4 A DETERMINANT APPROACH

At the conclusion of this chapter we characterize simplex cycles of (4.1.4) with the aid of certain inequality systems which the subdeterminants of Y must satisfy (cf. Theorem 4.4.2.1 and Corollary 4.4.2.4).

4.4.1 TERMS AND FOUNDATIONS

Let the reduced linear optimization problem (4.1.4) be given. For *arbitrary*[133] indices $j_1, \ldots, j_\sigma, \nu \in \{1, \ldots, n+\sigma\}$ we define $\sigma \times \sigma$- and $(\sigma+1) \times (\sigma+1)$-subdeterminants of

$$\begin{pmatrix} \bar{Y} \\ \hline -\bar{d} \end{pmatrix} = \begin{pmatrix} y_{1,1}, & \ldots, & y_{1,n+\sigma} \\ \vdots & & \vdots \\ y_{\sigma,1}, & \ldots, & y_{\sigma,n+\sigma} \\ -d_1, & \ldots, & -d_{n+\sigma} \end{pmatrix}$$

as follows:

$$D_{j_1,\ldots,j_\sigma} := \begin{vmatrix} y_{1,j_1}, & \ldots, & y_{1,j_\sigma} \\ \vdots & & \vdots \\ y_{\sigma,j_1}, & \ldots, & y_{\sigma,j_\sigma} \end{vmatrix} \tag{4.4.1.1}$$

and

$$\bar{D}_{j_1,\ldots,j_\sigma,\nu} := \begin{vmatrix} y_{1,j_1}, & \ldots, & y_{1,j_\sigma}, & y_{1,\nu} \\ \vdots & & \vdots & \vdots \\ y_{\sigma,j_1}, & \ldots, & y_{\sigma,j_\sigma}, & y_{\sigma,\nu} \\ -d_{j_1}, & \ldots, & -d_{j_\sigma}, & -d_\nu \end{vmatrix} \tag{4.4.1.2}$$

For indices $j_1, \ldots, j_\sigma$ with $1 \leq j_1 < \ldots < j_\sigma \leq n+\sigma$ let $B_{j_1,\ldots,j_\sigma} \in \mathbb{R}^{\sigma\times\sigma}$ denote the basis associated with the index set $\{j_1, \ldots, j_\sigma\}$ and let

---

[133] This means in particular that the indices $j_1,\ldots,j_\sigma,\nu$ need not be pairwise different and are not necessarily arranged in increasing order.

$$T_{j_1,\dots,j_\sigma} := \begin{pmatrix} B^{-1}_{j_1,\dots,j_\sigma}\bar{Y} \\ \hline (d_{j_1},\dots,d_{j_\sigma})B^{-1}_{j_1,\dots,j_\sigma}\bar{Y} - \bar{d}^T \end{pmatrix} \in \mathbb{R}^{(\sigma+1)\times(n+\sigma)} \tag{4.4.1.3}$$

denote the corresponding tableau (cf. Lemma 4.3.1.1).

Now the tableau $T_{j_1,\dots,j_\sigma}$ can be represented as follows:

Theorem 4.4.1.1:[134]

The tableau $T_{j_1,\dots,j_\sigma}$ $(1 \le j_1 < \dots < j_\sigma \le n+\sigma)$ has the form[135]

$$T_{j_1,\dots,j_\sigma} := \frac{1}{D_{j_1,\dots,j_\sigma}} \cdot \begin{pmatrix} D_{1,j_2,\dots,j_\sigma}, & D_{2,j_2,\dots,j_\sigma}, & \dots, & D_{n+\sigma,j_2,\dots,j_\sigma} \\ D_{j_1,1,\dots,j_\sigma}, & D_{j_1,2,\dots,j_\sigma}, & \dots, & D_{j_1,n+\sigma,\dots,j_\sigma} \\ \vdots & \vdots & & \vdots \\ D_{j_1,\dots,j_{\sigma-1},1}, & D_{j_1,\dots,j_{\sigma-1},2}, & \dots, & D_{j_1,\dots,j_{\sigma-1},n+\sigma} \\ \bar{D}_{j_1,\dots,j_\sigma,1}, & \bar{D}_{j_1,\dots,j_\sigma,2}, & \dots, & \bar{D}_{j_1,\dots,j_\sigma,n+\sigma} \end{pmatrix}$$

Proof: a) According to Cramer's rule (cf. e.g. Kowalsky (1975:96)), the j-th column y^j of $\bar{Y}$ can be represented as

$$B^{-1}_{j_1,\dots,j_\sigma} y^j = \frac{1}{D_{j_1,\dots,j_\sigma}} \begin{pmatrix} D_{j,j_2,\dots,j_\sigma} \\ D_{j_1,j,\dots,j_\sigma} \\ \vdots \\ D_{j_1,\dots,j_{\sigma-1},j} \end{pmatrix} \tag{4.4.1.4}$$

$(j = 1,\dots,n+\sigma)$.

b) Expansion of the determinant $\bar{D}_{j_1,\dots,j_\sigma,\nu}$ by the last row yields (cf.(4.4.1.2)):

---

134 Though this statement represents an elementary result of linear optimization – it enables us to represent an arbitrary simplex tableau *explicitly* using elements of the initial tableau – it is not to be found in standard literature (cf. e.g. Dantzig (1966), Hadley (1975), Vogel (1967), Yudin/Gol'shtein (1965)).

135 For the element in the i-th row and the j-th column the index j has been substituted for the i-th index of the sequence $j_1,\dots,j_\sigma$ $(i=1,\dots,\sigma; j=1,\dots,n+\sigma)$.

$$\left.\begin{aligned}\bar{D}_{j_1,\ldots,j_\sigma,\nu} =&(-1)^\sigma(-d_{j_1}D_{j_2,\ldots,j_\sigma,\nu} + d_{j_2}D_{j_1,j_3,\ldots,j_\sigma,\nu}\\ &- d_{j_3}D_{j_1,j_2,j_4,\ldots,j_\sigma,\nu} \pm \ldots\\ &(-1)^\sigma d_{j_\sigma}D_{j_1,\ldots,j_{\sigma-1},\nu}\\ &- (-1)^\sigma d_\nu D_{j_1,\ldots,j_\sigma})\end{aligned}\right\} \quad (4.4.1.5)$$

Since exchanging two columns of a matrix changes the sign of its determinant, the following holds:

$$D_{j_1,\ldots,j_{k-1},j_{k+1},\ldots,j_\sigma,\nu} = \begin{cases}(-1)^\sigma D_{j_1,\ldots,j_{k-1},\nu,j_{k+1},\ldots,j_\sigma} & \text{if } k \text{ is odd}\\ -(-1)^\sigma D_{j_1,\ldots,j_{k-1},\nu,j_{k+1},\ldots,j_\sigma} & \text{if } k \text{ is even.}\end{cases}$$

$$(4.4.1.6)$$

Using this we can transform (4.4.1.5) as follows:

$$\begin{aligned}\bar{D}_{j_1,\ldots,j_\sigma,\nu} = \quad & (-1)^{2\sigma}(d_{j_1}D_{\nu,j_2,\ldots,j_\sigma} + d_{j_2}D_{j_1,\nu,j_3,\ldots,j_\sigma} + \ldots\\ & +d_{j_\sigma}D_{j_1,\ldots,j_{\sigma-1},\nu} - d_\nu D_{j_1,\ldots,j_\sigma})\end{aligned}$$

$$= (d_{j_1},\ldots,d_{j_\sigma}) \cdot \begin{pmatrix} D_{\nu,j_2,\ldots,j_\sigma}\\ D_{j_1,\nu,\ldots,j_\sigma}\\ \vdots\\ D_{j_1,\ldots,j_{\sigma-1},\nu}\end{pmatrix} - d_\nu D_{j_1,\ldots,j_\sigma} \quad (4.4.1.7)$$

From this and (4.4.1.4) follows

$$\frac{\bar{D}_{j_1,\ldots,j_\sigma,\nu}}{D_{j_1,\ldots,j_\sigma}} = (d_{j_1},\ldots,d_{j_\sigma})B^{-1}_{j_1,\ldots,j_\sigma}y^\nu - d_\nu \quad (4.4.1.8)$$

Now the assertion follows from (4.4.1.3), (4.4.1.4) and (4.4.1.8). •

Example 4.4.1.2:

Let the reduced problem (4.1.6) be given. The tableau $T_{2,3}$ has the form

$$T_{2,3} = \frac{1}{D_{2,3}} \begin{pmatrix} D_{1,3}, & D_{2,3}, & D_{3,3}, & D_{4,3}, & D_{5,3}, & D_{6,3} \\ D_{2,1}, & D_{2,2}, & D_{2,3}, & D_{2,4}, & D_{2,5}, & D_{2,6} \\ \bar{D}_{2,3,1}, & \bar{D}_{2,3,2}, & \bar{D}_{2,3,3}, & \bar{D}_{2,3,4}, & \bar{D}_{2,3,5}, & \bar{D}_{2,3,6} \end{pmatrix} \quad (4.4.1.9).$$

From

$$\begin{pmatrix} \bar{Y} \\ \hline -\bar{d} \end{pmatrix} = \begin{pmatrix} \frac{1}{4}, & -60, & -\frac{1}{25}, & 9, & 1, & 0 \\ \frac{1}{2}, & -90, & -\frac{1}{50}, & 3, & 0, & 1 \\ -\frac{3}{4}, & 150, & -\frac{1}{50}, & 6, & 0, & 0 \end{pmatrix}$$

follows e.g.

$$D_{2,3} = \begin{vmatrix} -60 & -\frac{1}{25} \\ -90 & -\frac{1}{50} \end{vmatrix} = -\frac{12}{5}, D_{5,3} = \begin{vmatrix} 1 & -\frac{1}{25} \\ 0 & -\frac{1}{50} \end{vmatrix} = -\frac{1}{50},$$

$$\bar{D}_{2,3,6} = \begin{vmatrix} -60 & -\frac{1}{25} & 0 \\ -90 & -\frac{1}{50} & 1 \\ 150 & -\frac{1}{50} & 0 \end{vmatrix} = -\frac{36}{5}.$$

Inserting all determinants in (4.4.1.9) yields

$$T_{2,3} = -\frac{5}{12} \cdot \begin{pmatrix} \frac{3}{200} & -\frac{12}{5} & 0 & -\frac{3}{50} & -\frac{1}{50} & \frac{1}{25} \\ -\frac{15}{2} & 0 & -\frac{12}{5} & 630 & 90 & -60 \\ -\frac{3}{5} & 0 & 0 & \frac{36}{5} & \frac{24}{5} & -\frac{36}{5} \end{pmatrix}.$$

After exchanging rows this tableau coincides with the respective tableau in Tab. 4.1.2.

Taking the pivoting rules of Section 4.3.1 as a basis we can state

Theorem 4.4.1.3:

Let the tableau $T_{j_1,\ldots,j_\sigma}$ of the reduced problem (4.1.4) be given $(1 \leq j_1 < \ldots < j_\sigma \leq n+\sigma)$.

A simplex step, substituting the index ν for j_μ, can be performed if and only if the following conditions are satisfied $(1 \leq \mu \leq \sigma; \nu \in \{1,\ldots,n+\sigma\} \backslash \{j_1,\ldots,j_\sigma\})$:

a) $\frac{D_{j_1,\ldots,j_{\mu-1},\nu,j_{\mu+1},\ldots,j_\sigma}}{D_{j_1,\ldots,j_\sigma}} > 0$

b) $\frac{\bar{D}_{j_1,\ldots,j_\sigma,\nu}}{\bar{D}_{j_1,\ldots,j_\sigma}} < 0$

c) $\frac{\bar{D}_{j_1,\ldots,j_\sigma,\nu}}{D_{j_1,\ldots,j_\sigma}} \leq \frac{\bar{D}_{j_1,\ldots,j_\sigma,\bar{\nu}}}{D_{j_1,\ldots,j_\sigma}} \quad \forall \bar{\nu} = 1,\ldots,n+\sigma$

Proof: The assertion follows immediately from the representation of the tableau $T_{j_1,\ldots,j_\sigma}$ in Theorem 4.2.3.1.1. •

4.4.2 CHARACTERIZATION OF SIMPLEX CYCLES BY MEANS OF DETERMINANT INEQUALITY SYSTEMS

First of all we characterize simplex cycles without using concepts of the theory of degeneracy graphs.

Theorem 4.4.2.1:

Given the reduced problem (4.1.4) and a sequence of index sets

$$\left.\begin{array}{ll} & C = (I_1,\ldots,I_k,I_1) \\ \text{with} & \\ & I_\nu = \{j_1^{(\nu)},\ldots,j_\sigma^{(\nu)}\} \subset \{1,\ldots,n+\sigma\} \\ \text{and} & \\ & |I_{\nu+1}\backslash I_\nu| = 1 \quad \text{for} \quad \nu = 1,\ldots,k \\ & (I_{k+1} := I_1) \end{array}\right\} \qquad (4.4.2.1)$$

Let the indices j_ν^+ and $j_{\mu_\nu}^{(\nu)}$ be (uniquely) defined by (cf. Theorem 4.4.1.3):

$$j_\nu^+ \in I_{\nu+1}\backslash I_\nu,$$

$$j_{\mu_\nu}^{(\nu)} \in I_\nu\backslash I_{\nu+1}.$$

Then the above sequence C is a simplex cycle of (4.1.4) if and only if the following determinant inequality system is satisfied (cf. (4.4.1.1), (4.4.1.2)):

a) $D_{j_1^{(\nu)},\ldots,j_\sigma^{(\nu)}} \neq 0 \quad \forall \nu = 1,\ldots,k$

b) $\dfrac{D_{j_1^{(\nu)},\ldots,j_{\mu_\nu-1}^{(\nu)},j_\nu^+,j_{\mu_\nu+1}^{(\nu)},\ldots,j_\sigma^{(\nu)}}}{D_{j_1^{(\nu)},\ldots,j_\sigma^{(\nu)}}} > 0 \quad \forall \nu = 1,\ldots k$

c) $\dfrac{\bar{D}_{j_1^{(\nu)},\ldots,j_\sigma^{(\nu)},j_\nu^+}}{D_{j_1^{(\nu)},\ldots,j_\sigma^{(\nu)}}} < 0 \quad \forall \nu = 1,\ldots,k$

d) $\dfrac{\bar{D}_{j_1^{(\nu)},\ldots,j_\sigma^{(\nu)},j_\nu^+}}{D_{j_1^{(\nu)},\ldots,j_\sigma^{(\nu)}}} \leq \dfrac{\bar{D}_{j_1^{(\nu)},\ldots,j_\sigma^{(\nu)},\bar{\nu}}}{D_{j_1^{(\nu)},\ldots,j_\sigma^{(\nu)}}} \qquad \forall \nu = 1,\ldots,k; \forall \bar{\nu} = 1,\ldots,n+\sigma$

Proof: The condition a) means that the index sets of C correspond to bases of (4.1.4). Obviously C is a simplex cycle of (4.1.4) if and only if k simplex steps can be performed, where in the ν-th step j_ν^+ is substituted for $j_{\mu_\nu}^{(\nu)}$ $(\nu = 1,\ldots,k)$. Now the assertion follows from Theorem 4.4.1.3, since the conditions a) – c) there correspond to the above conditions b) – d). •

Theorem 4.4.2.1 is fundamental for the procedures in Chapter 5 which construct cycling examples. We will illustrate it by means of the special case with $\sigma = 2$ and $n = 4$:

Example 4.4.2.2:
Let the reduced problem (4.1.4) be of the special form $(\sigma = 2, n = 4)$:

$$\left.\begin{array}{llllllll} \max z = & d_1x_1 & +\ldots & +d_4x_4 & & & & \\ \text{s.t.} & & & & & & & \\ & y_{1,1}x_1 & +\ldots & +y_{1,4}x_4 & +x_5 & & & =0 \\ & y_{2,1}x_1 & +\ldots & +y_{2,4}x_4 & & & +x_6 & =0 \\ & & & & & & x_1,\ldots,x_6 & \geq 0 \end{array}\right\} \quad (4.4.2.2)$$

and let the sequence C in (4.4.2.1) be of the form $(k = 6)$:

$$C = (\{1,2\},\{2,3\},\{3,4\},\{4,5\},\{5,6\},\{1,6\},\{1,2\}). \qquad (4.4.2.3)$$

Theorem 4.4.2.1 implies that C is a simplex cycle if and only if the following system is satisfied:

$$D_{1,2}, D_{2,3}, D_{3,4}, D_{4,5}, D_{5,6}, D_{1,6} \neq 0 \tag{4.4.2.4}$$

$$\frac{D_{3,2}}{D_{1,2}}, \frac{D_{4,3}}{D_{2,3}}, \frac{D_{5,4}}{D_{3,4}}, \frac{D_{6,5}}{D_{4,5}}, \frac{D_{1,6}}{D_{5,6}}, \frac{D_{1,2}}{D_{1,6}} > 0 \tag{4.4.2.5}$$

$$\frac{\bar{D}_{1,2,3}}{D_{1,2}}, \frac{\bar{D}_{2,3,4}}{D_{2,3}}, \frac{\bar{D}_{3,4,5}}{D_{3,4}}, \frac{\bar{D}_{4,5,6}}{D_{4,5}}, \frac{\bar{D}_{5,6,1}}{D_{5,6}}, \frac{\bar{D}_{1,6,2}}{D_{1,6}} < 0 \tag{4.4.2.6}$$

$$\left.\begin{array}{lcll} \frac{\bar{D}_{1,2,3}}{D_{1,2}} & \leq & \frac{\bar{D}_{1,2,\bar{\nu}}}{D_{1,2}} & \forall \bar{\nu} = 1,\ldots,6 \\ \frac{\bar{D}_{2,3,4}}{D_{2,3}} & \leq & \frac{\bar{D}_{2,3,\bar{\nu}}}{D_{2,3}} & \forall \bar{\nu} = 1,\ldots,6 \\ \frac{\bar{D}_{3,4,5}}{D_{3,4}} & \leq & \frac{\bar{D}_{3,4,\bar{\nu}}}{D_{3,4}} & \forall \bar{\nu} = 1,\ldots,6 \\ \frac{\bar{D}_{4,5,6}}{D_{4,5}} & \leq & \frac{\bar{D}_{4,5,\bar{\nu}}}{D_{4,5}} & \forall \bar{\nu} = 1,\ldots,6 \\ \frac{\bar{D}_{5,6,1}}{D_{5,6}} & \leq & \frac{\bar{D}_{5,6,\bar{\nu}}}{D_{5,6}} & \forall \bar{\nu} = 1,\ldots,6 \\ \frac{\bar{D}_{1,6,2}}{D_{1,6}} & \leq & \frac{\bar{D}_{1,6,\bar{\nu}}}{D_{1,6}} & \forall \bar{\nu} = 1,\ldots,6 \end{array}\right\} \tag{4.4.2.7}$$

After certain redundant inequalities have been omitted, the system (4.4.2.4) - (4.4.2.7) can be transformed into :[136]

$$D_{1,2}, D_{3,2}, D_{3,4}, D_{5,4}, D_{1,6} > 0 \tag{4.4.2.8}$$

and

---

[136] Note that $D_{5,6}=1$ (cf. (4.4.2.2)) and that the determinants $D_{i,j}$, $\bar{D}_{i,j,\nu}$ change their sign when two indices are exchanged.

$$\left.\begin{array}{ll}
\bar{D}_{1,2,3} & \begin{cases} < & 0 \\ \leq & \bar{D}_{1,2,4}, \bar{D}_{1,2,5}, \bar{D}_{1,2,6} \end{cases} \\
\bar{D}_{3,2,4} & \begin{cases} < & 0 \\ \leq & \bar{D}_{3,2,5}, \bar{D}_{3,2,6}, \bar{D}_{3,2,1} \end{cases} \\
\bar{D}_{3,4,5} & \begin{cases} < & 0 \\ \leq & \bar{D}_{3,4,6}, \bar{D}_{3,4,1}, \bar{D}_{3,4,2} \end{cases} \\
\bar{D}_{5,4,6} & \begin{cases} < & 0 \\ \leq & \bar{D}_{5,4,1}, \bar{D}_{5,4,2}, \bar{D}_{5,4,3} \end{cases} \\
\bar{D}_{5,6,1} & \begin{cases} < & 0 \\ \leq & \bar{D}_{5,6,2}, \bar{D}_{5,6,3}, \bar{D}_{5,6,4} \end{cases} \\
\bar{D}_{1,6,2} & \begin{cases} < & 0 \\ \leq & \bar{D}_{1,6,3}, \bar{D}_{1,6,4}, \bar{D}_{1,6,5} \end{cases}
\end{array}\right\} \qquad (4.4.2.9)^{137}$$

where we obtain (4.4.2.8) from (4.4.2.4), (4.4.2.5) and (4.4.2.9.) from (4.4.2.6), (4.4.2.7).

Example 4.4.2.3:

Given the problem (4.1.6) and the cycle C in (4.4.2.3). The determinants in (4.4.2.8), (4.4.2.9) are as follows (cf. Example 4.4.1.2):

$$\begin{array}{llll}
D_{1,2} = \frac{15}{2}, & D_{3,2} = \frac{12}{5}, & D_{3,4} = \frac{3}{50} & D_{5,4} = 3 \\
D_{1,6} = \frac{1}{4} & & & \\
\bar{D}_{1,2,3} = -\frac{3}{5}, & \bar{D}_{1,2,4} = 135, & \bar{D}_{1,2,5} = \frac{15}{2}, & \bar{D}_{1,2,6} = \frac{15}{2}, \\
\bar{D}_{3,2,4} = -\frac{36}{5}, & \bar{D}_{3,2,5} = -\frac{24}{5}, & \bar{D}_{3,2,6} = \frac{36}{5}, & \bar{D}_{3,2,1} = \frac{3}{5}, \\
\bar{D}_{3,4,5} = -\frac{3}{50}, & \bar{D}_{3,4,6} = \frac{3}{50}, & \bar{D}_{3,4,1} = -\frac{3}{100}, & \bar{D}_{3,4,2} = \frac{36}{5}, \\
\bar{D}_{5,4,6} = -6, & \bar{D}_{5,4,1} = -\frac{21}{4}, & \bar{D}_{5,4,2} = 990, & \bar{D}_{5,4,3} = \frac{3}{50}, \\
\bar{D}_{5,6,1} = -\frac{3}{4} & \bar{D}_{5,6,2} = 150, & \bar{D}_{5,6,3} = -\frac{1}{50}, & \bar{D}_{5,6,4} = 6 \\
\bar{D}_{1,6,2} = -\frac{15}{2}, & \bar{D}_{1,6,3} = -\frac{7}{200}, & \bar{D}_{1,6,4} = \frac{33}{4}, & \bar{D}_{1,6,5} = \frac{3}{4}
\end{array}$$

The above determinants satisfy the systems (4.4.2.8), (4.4.2.9), i.e. according to Theorem 4.4.2.1 the cycle C in(4.4.2.3) is a simplex cycle of (4.1.6).

---

137 If we admit arbitrary columns with negative relative cost coefficients as pivot columns, the inequalities of type " $\leq$ " have to be dropped (cf. Section 4.2.2). The subsystem obtained thereby is equivalent to the system (9.48)-(9.58) in Yudin/Gol'shtein (1965:245).

The formulation of Theorem 4.4.2.1 can be simplified as follows using the terminology of degeneracy graphs:

Corollary 4.4.2.4:

Let the reduced problem (4.1.4) and a cycle $C = (I_1, \dots, I_k, I_1)$ of the positive degeneracy graph G_Y^+ be given ($I_\nu = \{j_1^{(\nu)}, \dots, j_\sigma^{(\nu)}\}$ for $\nu = 1, \dots, k$; cf. Theorem 4.4.2.1).

The cycle C is a simplex cycle of (4.1.4) if and only if the conditions c) and d) in Theorem 4.4.2.1 are satisfied.

Proof: The left-hand side in the condition b) of Theorem 4.4.2.1 consists of the pivot element of the ν-th tableau which effects the transition to the $(\nu+1)$-th tableau (cf. Theorem 4.4.1.1). Therefore this condition means that the bases (tableaux) associated with I_ν and $I_{\nu+1}$ are transformable each into the other by a positive pivot step.

Hence the conditions a) and b) in Theorem 4.4.2.1 mean that the sequence C in (4.4.2.1) is a cycle of G_Y^+ (cf. the proof of Theorem 4.4.2.1). •

SUMMARY OF CHAPTER 4

Many recent articles on simplex cycling point out that this phenomenon considerably impairs the efficiency of the simplex algorithm, particularly in solving large-scale problems. Any degenerate vertex passed during the simplex procedure may cause a loss in efficiency.

Though this problem was already known in the fifties, and although a multitude of anticycling rules exists (more and more are appearing in the literature), it has not been possible to date to reveal the reasons for simplex cycling completely. Thus in Chapter 4 degeneracy graphs have been used as an instrument to clarify the facts.

Stated simply, the following results are derived: The tableau sequence of a cycling example can always be represented as a cycle C of the positive degeneracy graph G_Y^+ of a reduced linear optimization problem (RLOP). On this basis degeneracy graphs can be applied for explaining simplex cycling. The question is, which structural properties G_Y^+ or the matrix of coefficients of (RLOP) must have, such that C is a simplex cycle. In answering this question we introduce the following three approaches:

1) ***A pure graph theoretical approach***
   We show that a cycle C is a simplex cycle if and only if it can be enlarged to a "star-shaped" graph contained in the LP-degeneracy graph $G_{Y,d}$.
2) ***Geometrically motivated approaches***
   The main assertion is (in simplified form): The cycle C is a simplex cycle if and only if the column vectors of the matrix of coefficients of (RLOP) are contained in the induced point set P_C.
3) ***A determinant approach***
   We show that C is a simplex cycle if and only if the coefficients of (RLOP) satisfy a certain determinant inequality system.

On the one hand the results of Chapter 4 are of theoretical interest, since simplex cycling is studied and explained from different points of view. Moreover, these statements can be used to construct cycling examples which enable us to clarify questions of practical interest (cf. Section 5.1). In particular, the results using the above approach 1) represent the theoretical foundation for an algorithm detecting simplex cycles of a (reduced) linear optimization problem. Such an algorithm would permit the decision by means of structural properties of the LP-degeneracy graph (or the matrix of coefficients of (RLOP)), of whether simplex cycling is possible at all. However, additional research is necessary in this field.

5. PROCEDURES FOR CONSTRUCTING CYCLING EXAMPLES

The concepts presented in Chapter 4 are intended to explain simplex cycling theoretically. In order to study this phenomenon from a practical point of view, a large number of cycling examples with dimensions as large as possible is required (cf. Section 5.1). However, only few specific cycling examples with small dimensions have been published in the past[138], examples which only serve merely to demonstrate the possibility of simplex cycling. A systematic method for the construction of cycling in linear optimization is not known to date.[139]

For this reason we present procedures for the construction of an arbitrary number of cycling examples (also with larger dimensions, cf. Sections 5.2.3, 5.2.4 and 5.2.6).

First of all we present applications of constructed cycling examples (Section 5.1) and subsequently certain successive procedures for constructing reduced cycling examples (Section 5.2). In conclusion we show how general cycling examples (i.e. those in which the right-hand side differs from zero) are obtained (Section 5.3).

5.1 ON THE PRACTICAL USE OF CONSTRUCTED CYCLING EXAMPLES

In the first years after the simplex method was published by Dantzig (1951), cycling was considered to be a pure theoretical problem.[140] Cycling in practice was probably mentioned for the first time by Wolfe (1963:205). Later Benichou et al. (1977:292), Gass (1979:850) and Kotiah/Steinberg (1977; 1978) discovered instances of cycling in practice (cf. also McKeown (1978:355)). Occasionally cycling has been

138 Cf. Beale (1955), Cunningham (1979), Gassner (1964), Hoffmann (1953), Marshall/Suurballe (1969), Roos (1984), Solow (1984:178).

139 A rudimentary method for generating cycling examples for network problems is described in Cunningham/Klincewicz (1983:188).

140 Cf. e.g. Beale (1955:269), Dantzig/Orden/Wolfe (1955:184).

observed in applications but has not been reported in literature (cf. Kotiah/Steinberg (1978:375)).[141]

In spite of the above observations and although more and more anti-cycling rules have been developed[142], the practical meaning of simplex cycling has been controversial till now.[143] The following assumptions have been stated:

- Whenever cycling occurs in practice, rounding errors perturb the problem, causing the algorithm to exit from the cycle "automatically" after a number of iterations (cf. Cooper/Steinberg (1974), Kotiah/Steinberg (1977:111)).
- Solving practical linear optimization problems using "professional" programs requires a distinction between classical cycling and computer cycling.[144] The former means cycling under "rational arithmetics", where *absolutely exact computations* are performed; computer cycling occurs in applying certain program packages, which include a modified version of the simplex algorithm and use standardized *rounding procedures.*[145] It is assumed that cycling observed in practice is not always due to classical cycling but is often exclusively caused by rounding errors (cf. Gass (1979:850; 1985:183)).

Constructed cycling examples enable us to perform representative tests in order to prove the above assertions. It must be proved for different program packages whether or after how many iterations a cycle is exited

---

141 Problems related to simplex cycling which are important for practice, are "stalling" ("long" sequences of bases, associated with a degenerate vertex; cf. Cunningham (1979), Hattersley/Wilson (1988:136)) and "near cycling" ("cycling of variables in and out of the basis" with a "slow" improvement of the objective function value; cf. Thompson et al. (1966:601)).

142 Cf. the introductory representations in Chapter 4.

143 Cf. e.g. Fletcher (1987:178), Ecker/Kupferschmid (1988:53) and Majthay (1981).

144 Cf. Gass (1979; 1985:183) and Ecker/Kupferschmid (1988).

145 Note that whether computer cycling occurs or not depends on the respective program package (Telgen (1980:8).

due to rounding errors, or whether computer cycling occurs. Telgen (1980:11) stated another assumption on the occurrence of cycling:

- Simplex cycling is improbable if the constraints are a minimal representation of the feasible solution set, i.e. if all constraints are nonredundant.

We can test this assumption as well by means of numerous constructed cycling examples if we investigate their solution sets with respect to redundancy.

Finally, constructed cycling examples can be useful in empirical tests of the efficiency of anticycling rules.[146]
Quite recently Hattersley/Wilson (1988: 135ff.) detected certain numerical difficulties in often-used anticycling rules. In the literature the following tests of the above kind are known:[147]

- Avis/Chvatal (1978) compared Bland's (1977) first anticycling rule ((γ) "smallest subscript rule") with two pivoting rules which do not exclude cycling:
  (α) "smallest relative cost rule" (cf. Section 4.3.1)
  (β) "rule of largest increase of the objective function".
  Generally speaking, rule (γ) required considerably more iterations than (α) and (β) in solving randomly generated problems (e.g. for 50×50 problems the number of iterations for (γ) was about four times as high as for (α)).
- Ciriná (1985) investigated (α), (γ) and two additional rules based on (γ). He applied them to linear problems generated by the random procedure of Avis/Chvátal (1978). The tests showed that Ciriná's rules are at least as good as rule (α) (in respect to the iteration number as well as to CPU-time).
- Finally Barr/Glover/Klingman (1977) compared their "alternating basis algorithm" for solving assignment problems (which

[146] Cf. the footnote to the introductory representations of Ch. 4.

[147] Moreover, a lot of practical tests (and theoretical investigations) exist for estimating the efficiency of pivoting rules, which have not been developed in order to avoid cycling but to reduce the number of iterations in solving randomly generated linear problems (cf. e.g. Abel (1987), Adler et al. (1986), Kuhn/Quandt (1963), Todd (1986, Section 4) and the survey of Shamir (1987)).

avoids simplex cycling in particular) with five other algorithms. They applied the procedures to 200×200 assignment problems.[148] The alternating basis algorithm proved to require about 25% fever iterations than the most efficient version of the network simplex method.

However, as a rule randomly generated linear optimization problems are not cycling examples. Thus it is doubtful whether the preceding investigations really permit statements about the "goodness" of anticycling rules. This can be proved by means of a test series in which anticycling rules are applied to a large set of constructed cycling examples. In these cases the computational effort per iteration and the number of iterations needed for exiting the degenerate vertex have to be tested.[149]

In order to decide whether an anticycling rule requires only a "small" number of iterations, i.e. whether it exits the degeneracy graph by a "short" path, the results on the diameter of degeneracy graphs can be used (cf. Theorem 3.2.3.2). If in a test it turns out that a rule exits a $\sigma \times n$-degeneracy graph after about $s = \min\{\sigma, n\}$ steps, the iteration number should be considered as "small", since possibly no shorter path exists for exiting the degeneracy graph.

5.2 SUCCESSIVE PROCEDURES FOR CONSTRUCTING CYCLING EXAMPLES

According to Theorem 4.4.2.1, a reduced linear optimization problem of the form

$$\left.\begin{aligned} &\max z = \bar{d}^T \bar{x} \\ &\text{s.t.} \\ &\bar{Y}\bar{x} = 0, \bar{x} \geq 0 \end{aligned}\right\} \tag{5.2.1}$$

(cf. (4.1.4)) is a cycling example with simplex cycle

$$C = (I_1, \ldots, I_k, I_1) \tag{5.2.2}$$

[148] Probably the assignment problems were generated randomly.

[149] A related problem is the estimation of the length of a lexicographic path (cf. Balinski et al. (1986) and Section 2.2.2b)).

(cf. (4.4.2.1)) if and only if the elements of $\bar{Y}$ and $\bar{d}$ satisfy the (nonlinear) determinant inequality system (a) – (d) in Theorem 4.4.2.1. Thus in principle any known solution method for nonlinear inequality systems could be applied for determining reduced[150] cycling examples. However, this involves considerable computational difficulties, since system (a) – (d) is very complex (cf. Appendix C).

In order to avoid such problems, we present in this section successive procedures for constructing cycling examples. Starting with the initial tableau of a given cycling example, we modify a row (column) or add another row (column), step by step. Since the elements outside of the respective row (column) remain unchanged, a *linear* system must be solved in each step (cf. Theorem 4.4.2.1).

We begin by illustrating the construction steps by means of numerical cycling examples with small dimensions (Sections 5.2.1 – 5.2.4).

Combined procedures are introduced in Section 5.2.5. Finally we discuss some questions arising in connection with their practical performance (Section 5.2.6).

5.2.1 MODIFICATION OF A ROW IN THE INITIAL TABLEAU

We now describe the construction step, modifying a row of the initial tableau of a given cycling example. Let the cycling example

$$\left.\begin{array}{lllllllr} \max z = & \frac{3}{4}x_1 & - 150x_2 & + \frac{1}{50}x_3 & - 6x_4 & & & \\ \text{s.t.} & & & & & & & \\ & \frac{1}{4}x_1 & - 60x_2 & - \frac{1}{25}x_3 & + 9x_4 & + x_5 & & = 0 \\ & \frac{1}{2}x_1 & - 90x_2 & - \frac{1}{50}x_3 & + 3x_4 & & + x_6 & = 0 \\ & & & & & & x_1,\ldots,x_6 & \geq 0 \end{array}\right\} \quad (5.2.1.1)$$

with the simplex cycle

$$C = (\{1,2\},\{2,3\},\{3,4\},\{4,5\},\{5,6\},\{1,6\},\{1,2\}) \quad (5.2.1.2)$$

[150] In Section 5.3 we show how general cycling examples can be obtained from the reduced ones.

be given (cf. Example 4.1.2). We will now modify the first row in the initial tableau of (5.2.1.1), i.e. we determine "new" coefficients $y_{1,1},\ldots,y_{1,4}$ such that

$$\left.\begin{array}{lllllllll}
\max z = & \frac{3}{4}x_1 - & 150x_2 + & \frac{1}{50}x_3 - & 6x_4 & & & & \\
\text{s.t.} & & & & & & & & \\
& y_{1,1}x_1 + & y_{1,2}x_2 + & y_{1,3}x_3 + & y_{1,4}x_4 + & x_5 & & & = 0 \\
& \frac{1}{2}x_1 - & 90x_2 - & \frac{1}{50}x_3 + & 3x_4 + & & x_6 & & = 0 \\
& & & & & & x_1,\ldots,x_6 & & \geq 0
\end{array}\right\} \tag{5.2.1.3}$$

is a cycling example with the simplex cycle C in (5.2.1.2). According to Theorem 4.4.2.1, we must solve the corresponding determinant inequality system

$$\left.\begin{array}{l}
D_{1,2}, D_{3,2}, D_{3,4}, D_{5,4}, D_{1,6} > 0 \\
\bar{D}_{1,2,3}\begin{cases} < 0 \\ \leq \bar{D}_{1,2,4}, \bar{D}_{1,2,5}, \bar{D}_{1,2,6} \end{cases} \\
\bar{D}_{3,2,4}\begin{cases} < 0 \\ \leq \bar{D}_{3,2,5}, \bar{D}_{3,2,6}, \bar{D}_{3,2,1} \end{cases} \\
\bar{D}_{3,4,5}\begin{cases} < 0 \\ \leq \bar{D}_{3,4,6}, \bar{D}_{3,4,1}, \bar{D}_{3,4,2} \end{cases} \\
\bar{D}_{5,4,6}\begin{cases} < 0 \\ \leq \bar{D}_{5,4,1}, \bar{D}_{5,4,2}, \bar{D}_{5,4,3} \end{cases} \\
\bar{D}_{5,6,1}\begin{cases} < 0 \\ \leq \bar{D}_{5,6,2}, \bar{D}_{5,6,3}, \bar{D}_{5,6,4} \end{cases} \\
\bar{D}_{1,6,2}\begin{cases} < 0 \\ \leq \bar{D}_{1,6,3}, \bar{D}_{1,6,4}, \bar{D}_{1,6,5} \end{cases}
\end{array}\right\} \tag{5.2.1.4}$$

Inserting the coefficients of (5.2.1.3) for the determinants in (5.2.1.4) (cf. Example 4.4.1.2 for the evaluation of determinants) yields the *linear* inequality system[151]

---

[151] Since system (5.2.1.4) or (5.2.1.5) is solved inserting the coefficients of the (reduced) cycling example of Beale (1955) (cf. Example 4.4.2.3), the inequalities in (5.2.1.5) are "automatically" satisfied if they contain none of the variables $y_{1,j}$.

$$
\left.
\begin{array}{r}
-90y_{1,1} - \dfrac{1}{2}y_{1,2} > 0 \\
\dfrac{1}{50}y_{1,2} - 90y_{1,3} > 0 \\
3y_{1,3} + \dfrac{1}{50}y_{1,4} > 0 \\
3 > 0 \\
y_{1,1} > 0 \\
\dfrac{24}{5}y_{1,1} + \dfrac{1}{40}y_{1,2} + \dfrac{15}{2}y_{1,3} \begin{cases} < 0 \\ \leq -990y_{1,1} - \frac{21}{4}y_{1,2} + \frac{15}{2}y_{1,4} \\ \leq \frac{15}{2} \\ \leq -150y_{1,1} - \frac{3}{4}y_{1,2} \end{cases} \\
\dfrac{3}{50}y_{1,2} - 990y_{1,3} - \dfrac{24}{5}y_{1,4} \begin{cases} < 0 \\ \leq -\frac{24}{5} \\ \leq -\frac{1}{50}y_{1,2} - 150y_{1,3} \\ \leq -\frac{24}{5}y_{1,1} - \frac{1}{40}y_{1,2} - \frac{15}{2}y_{1,3} \end{cases} \\
-\dfrac{3}{50} \begin{cases} < 0 \\ \leq -6y_{1,3} - \frac{1}{50}y_{1,4} \\ \leq -\frac{3}{50}y_{1,1} - \frac{21}{4}y_{1,3} - \frac{1}{40}y_{1,4} \\ \leq -\frac{3}{50}y_{1,2} + 990y_{1,3} + \frac{24}{5}y_{1,4} \end{cases} \\
-6 \begin{cases} < 0 \\ \leq -\frac{21}{4} \\ \leq 990 \\ \leq \frac{3}{50} \end{cases} \\
-\dfrac{3}{4} \begin{cases} < 0 \\ \leq 150 \\ \leq -\frac{1}{50} \\ \leq 6 \end{cases} \\
150y_{1,1} + \dfrac{3}{4}y_{1,2} \begin{cases} < 0 \\ \leq -\frac{1}{50}y_{1,1} + \frac{3}{4}y_{1,3} \\ \leq 6y_{1,1} + \frac{3}{4}y_{1,4} \\ \leq \frac{3}{4} \end{cases}
\end{array}
\right\}
\tag{5.2.1.5}
$$

Transforming (5.2.1.5) into the "normal form" of a linear inequality system yields (after omitting certain redundant inequalities and the inequalities without variables $y_{1,j}$):

$$\left.\begin{array}{rrrrcrr}
90y_{1,1} + & \frac{1}{2}y_{1,2} & & & < & 0 & (1) \\
 & -\frac{1}{50}y_{1,2} + & 90y_{1,3} & & < & 0 & (2) \\
 & & -3y_{1,3} - & \frac{1}{50}y_{1,4} & < & 0 & (3) \\
-y_{1,1} & & & & < & 0 & (4) \\
\frac{24}{5}y_{1,1} + & \frac{1}{40}y_{1,2} + & \frac{15}{2}y_{1,3} & & < & 0 & (5) \\
994,8y_{1,1} + & \frac{211}{40}y_{1,2} + & \frac{15}{2}y_{1,3} - & \frac{15}{2}y_{1,4} & \leq & 0 & (6) \\
154,8y_{1,1} + & \frac{31}{40}y_{1,2} + & \frac{15}{2}y_{1,3} & & \leq & 0 & (7) \\
 & \frac{3}{50}y_{1,2} - & 990y_{1,3} - & \frac{24}{5}y_{1,4} & \leq & -\frac{24}{5} & (8) \\
 & \frac{2}{25}y_{1,2} - & 840y_{1,3} - & \frac{24}{5}y_{1,4} & \leq & 0 & (9) \\
\frac{24}{5}y_{1,1} + & \frac{17}{200}y_{1,2} - & 982,5y_{1,3} - & \frac{24}{5}y_{1,4} & \leq & 0 & (10) \\
 & & 6y_{1,3} + & \frac{1}{50}y_{1,4} & \leq & \frac{3}{50} & (11) \\
\frac{3}{50}y_{1,1} + & & \frac{21}{4}y_{1,3} + & \frac{1}{40}y_{1,4} & \leq & \frac{3}{50} & (12) \\
150y_{1,1} + & \frac{3}{4}y_{1,2} & & & < & 0 & (13) \\
150,02y_{1,1} + & \frac{3}{4}y_{1,2} - & \frac{3}{4}y_{1,3} & & \leq & 0 & (14) \\
144y_{1,1} + & \frac{3}{4}y_{1,2} - & & \frac{3}{4}y_{1,4} & \leq & 0 & (15)
\end{array}\right\} \quad (5.2.1.6)$$

Hence the linear problem (5.2.1.3) is a cycling example with simplex cycle C in (5.2.1.2) if and only if the coefficients $y_{1,1}, \ldots, y_{1,4}$ satisfy system (5.2.1.6). A solution of (5.2.1.6) is[152]

$$(y_{1,1}, y_{1,2}, y_{1,3}, y_{1,4}) = \left(\frac{1}{600}, \frac{-11}{30}, \frac{-1}{12000}, \frac{203}{200}\right).$$

Inserting these values in (5.2.1.3), we obtain the following cycling example with modified first row:

---

[152] The solution was found using a program for solving linear inequality systems of F. Geue (cf. Section 5.2.6).

Tab. 5.2.1.1
Tableaux of the cycling example (5.2.1.7)
with simplex cycle C in (5.2.1.2)

	1	2	3	4	5	6
5	$\boxed{\frac{1}{600}}$	$-\frac{11}{30}$	$-\frac{1}{12000}$	$\frac{203}{200}$	1	0
6	$\frac{1}{2}$	-90	$-\frac{1}{50}$	3	0	1
Δz_j	$-\frac{3}{4}$	150	$-\frac{1}{50}$	6	0	0
1	1	-220	$-\frac{1}{20}$	609	600	0
6	0	$\boxed{20}$	$\frac{1}{200}$	$-\frac{603}{2}$	-300	1
Δz_j	0	-15	$-\frac{23}{400}$	$\frac{1851}{4}$	450	0
1	1	0	$\boxed{\frac{1}{200}}$	$-\frac{5415}{2}$	-2700	11
2	0	1	$\frac{1}{4000}$	$-\frac{603}{40}$	-15	$\frac{1}{20}$
Δz_j	0	0	$-\frac{43}{800}$	$\frac{1893}{8}$	225	$\frac{3}{4}$
3	200	0	1	-541500	-540000	2200
2	$-\frac{1}{20}$	1	0	$\boxed{\frac{1203}{10}}$	120	$-\frac{1}{2}$
Δz_j	$\frac{43}{4}$	0	0	-28869	-28800	119
3	$-\frac{10050}{401}$	$\frac{1805000}{401}$	1	0	$\boxed{\frac{60000}{401}}$	$-\frac{20300}{401}$
4	$-\frac{1}{2406}$	$\frac{10}{1203}$	0	1	$\frac{400}{401}$	$-\frac{5}{1203}$
Δz_j	$-\frac{2003}{1604}$	$\frac{96230}{401}$	0	0	$-\frac{1200}{401}$	$-\frac{396}{401}$
5	$-\frac{67}{400}$	$\frac{361}{12}$	$\frac{401}{60000}$	0	1	$-\frac{203}{600}$
4	$\frac{1}{6}$	-30	$-\frac{1}{150}$	1	0	$\boxed{\frac{1}{3}}$
Δz_j	$-\frac{7}{4}$	330	$\frac{1}{50}$	0	0	-2

The next pivot step generates the initial tableau.

Legend: Tableau associated with basis B

	column indices
indices of the basis B	$B^{-1}\bar{Y}$
Δz_j	$c_B^T B^{-1}\bar{Y} - \bar{d}$

(The right-hand side always equals zero and is therefore omitted.)

$$\left.\begin{array}{llllllll} \max z = & \frac{3}{4}x_1 & -150x_2 & +\frac{1}{50}x_3 & -6x_4 & & & \\ \text{s.t.} & & & & & & & \\ & \frac{1}{600}x_1 & -\frac{11}{30}x_2 & -\frac{1}{12000}x_3 & +\frac{203}{200}x_4 & +x_5 & & = 0 \\ & \frac{1}{2}x_1 & -90x_2 & -\frac{1}{50}x_3 & +3x_4 & & +x_6 & = 0 \\ & & & & & & x_1,\ldots,x_6 & \geq 0 \end{array}\right\} \quad (5.2.1.7)$$

The simplex cycle is C in (5.2.1.2). The associated tableaux are listed in Tab. 5.2.1.1.

5.2.2 MODIFICATION OF A COLUMN IN THE INITIAL TABLEAU

Instead of a row, a column can be modified in the initial tableau of a cycling example as well. We will illustrate this again by means of the cycling example (5.2.1.1) with simplex cycle C in (5.2.1.2).

We modify the first column of the initial tableau of (5.2.1.1), i.e. we determine "new" coefficients $y_{1,1}, y_{2,1}, d_1$ such that

$$\left.\begin{array}{llllllll} \max z = & d_1x_1 - & 150x_2 + & \frac{1}{50}x_3 - & 6x_4 & & & \\ \text{s.t.} & & & & & & & \\ & y_{1,1}x_1 + & 60x_2 + & \frac{1}{25}x_3 + & 9x_4 + & x_5 & & = 0 \\ & y_{2,1}x_1 - & 90x_2 - & \frac{1}{50}x_3 + & 3x_4 & & +x_6 & = 0 \\ & & & & & & x_1,\ldots,x_6 & \geq 0 \end{array}\right\} \quad (5.2.2.1)$$

is a cycling example with simplex cycle C in (5.2.1.2). According to Theorem 4.4.1.2, we must solve the corresponding determinant inequality system (5.2.1.4) (cf. Section 5.2.1). Inserting the coefficients of (5.2.2.1) for the determinants of (5.2.1.4) yields the linear system[153]

[153] In analogy to the footnote to (5.2.1.5), that inequalities are satisfied "automatically" which contain none of the variables $y_{1,1}, y_{2,1}, d_1$.

$$
\left.
\begin{array}{r}
-90y_{1,1} + 60y_{2,1} > 0 \\
\frac{12}{5} > 0 \\
\frac{3}{50} > 0 \\
3 > 0 \\
y_{1,1} > 0 \\
\frac{24}{5}y_{1,1} - \frac{36}{5}y_{2,1} + \frac{12}{5}d_1 \left\{ \begin{array}{l} < 0 \\ \leq -990y_{1,1} + 1710y_{2,1} - 630d_1 \\ \leq 150y_{2,1} - 90d_1 \\ \leq -150y_{1,1} + 60d_1 \end{array} \right. \\
-\frac{36}{5} \left\{ \begin{array}{l} < 0 \\ \leq -\frac{24}{5} \\ \leq \frac{36}{5} \\ \leq -\frac{24}{5}y_{1,1} + \frac{36}{5}y_{2,1} - \frac{12}{5}d_1 \end{array} \right. \\
-\frac{3}{50} \left\{ \begin{array}{l} < 0 \\ \leq \frac{3}{50} \\ \leq -\frac{3}{50}y_{1,1} + \frac{3}{50}y_{2,1} - \frac{3}{50}d_1 \\ \leq \frac{36}{5} \end{array} \right. \\
-6 \left\{ \begin{array}{l} < 0 \\ \leq -6y_{2,1} - 3d_1 \\ \leq 990 \\ \leq \frac{3}{50} \end{array} \right. \\
-d_1 \left\{ \begin{array}{l} < 0 \\ \leq 150 \\ \leq -\frac{1}{50} \\ \leq 6 \end{array} \right. \\
150y_{1,1} - 60d_1 \left\{ \begin{array}{l} < 0 \\ \leq -\frac{1}{50}y_{1,1} - \frac{1}{25}d_1 \\ \leq 6y_{1,1} + 9d_1 \\ \leq d_1 \end{array} \right.
\end{array}
\right\}
\qquad (5.2.2.2)
$$

Transforming (5.2.2.2) into the "normal form" of a linear inequality system yields (after omitting certain redundant inequalities and inequalities without any of the variables $y_{1,1}, y_{2,1}, d_1$):

$$\left.\begin{array}{rrrcrr}
90y_{1,1} - & 60y_{2,1} & & < & 0 & (1) \\
-y_{1,1} & & & < & 0 & (2) \\
4,8y_{1,1} - & 7,2y_{2,1} + & 2,4d_1 & < & 0 & (3) \\
994,8y_{1,1} - & 1717,2y_{2,1} + & 632,4d_1 & \leq & 0 & (4) \\
4,8y_{1,1} - & 157,2y_{2,1} + & 92,4d_1 & \leq & 0 & (5) \\
154,8y_{1,1} - & 7,2y_{2,1} - & 57,6d_1 & \leq & 0 & (6) \\
0,06y_{1,1} - & 0,06y_{2,1} + & 0,06d_1 & \leq & 0,06 & (7) \\
 & 6y_{2,1} + & 3d_1 & \leq & 6 & (8) \\
 & & -d_1 & \leq & -0,02 & (9) \\
150,02y_{1,1} - & & 59,96d_1 & \leq & 0 & (10)
\end{array}\right\} \quad (5.2.2.3)$$

(Note that three of the last four inequalities in (5.2.2.2) are redundant, since $y_{1,1}$ and d_1 must be positive.)

Hence the linear problem (5.2.2.1) is a cycling example with simplex cycle C in (5.2.1.2) if and only if the coefficients $y_{1,1}, y_{2,1}, d_1$ satisfy the system (5.2.2.3). We can solve (5.2.2.3) e.g. as follows:

Inserting $d_1 = \frac{1}{2}$ in (5.2.2.3)[154] leads to an inequality system with two variables, $y_{1,1}$ and $y_{1,2}$, which can be easily solved graphically.

A solution is e.g. $(y_{1,1}, y_{2,1}) = (\frac{1}{10}, \frac{7}{10})$ (cf. Fig. 5.2.2.1). It should be of interest that the inequalities (1), (3), (4), (6) and (7) are *strongly* redundant (cf. also Fig. 5.2.2.2).

A solution of (5.2.2.3) is $(y_{1,1}, y_{2,1}, d_1) = (\frac{1}{10}, \frac{7}{10}, \frac{1}{2})$.

Inserting these values in (5.2.2.1) yields the following cycling example with a modified left column:

$$\left.\begin{array}{lrrrrrrr}
\max z = & \frac{1}{2}x_1 - & 150x_2 + & \frac{1}{50}x_3 - & 6x_4 & & & \\
\text{s.t.} & & & & & & & \\
 & \frac{1}{10}x_1 - & 60x_2 - & \frac{1}{25}x_3 + & 9x_4 + & x_5 & & = 0 \\
 & \frac{7}{10}x_1 - & 90x_2 - & \frac{1}{50}x_3 + & 3x_4 & + & x_6 & = 0 \\
 & & & & & & x_1, \ldots, x_6 & \geq 0
\end{array}\right\} \quad (5.2.2.4)$$

The simplex cycle is C in (5.2.1.2). The associated tableaux are listed in Tab. 5.2.2.1.

[154] This is reasonable, since $\frac{1}{2}$ differs not much from the value $y_{1,1} = \frac{3}{4}$ in the underlying cycling example (5.2.1.1).

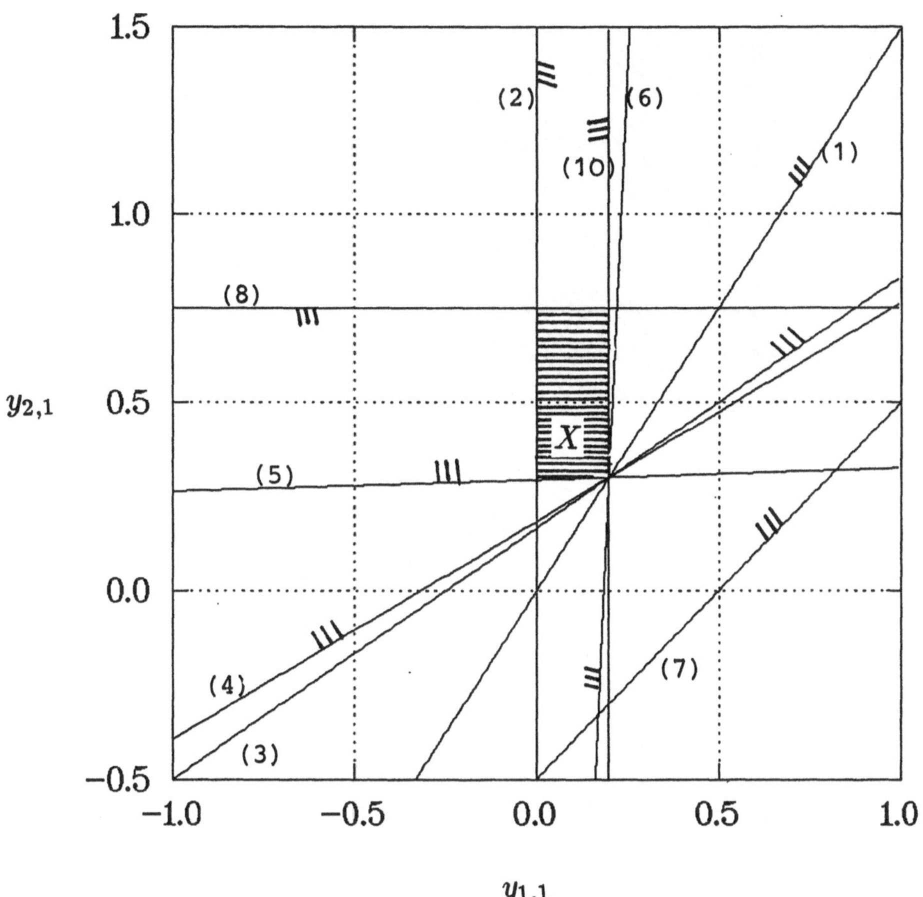

Legend: (i) – i-th boundary line
$(i = 1, \ldots, 10;\ i \neq 9)$
The hatchings indicate the positions of
the corresponding halfplanes.

Fig. 5.2.2.1
Representation of the solution set X of the system obtained from (5.2.2.3) for $d_1 = \frac{1}{2}$ (cf. also Fig. 5.2.2.2)

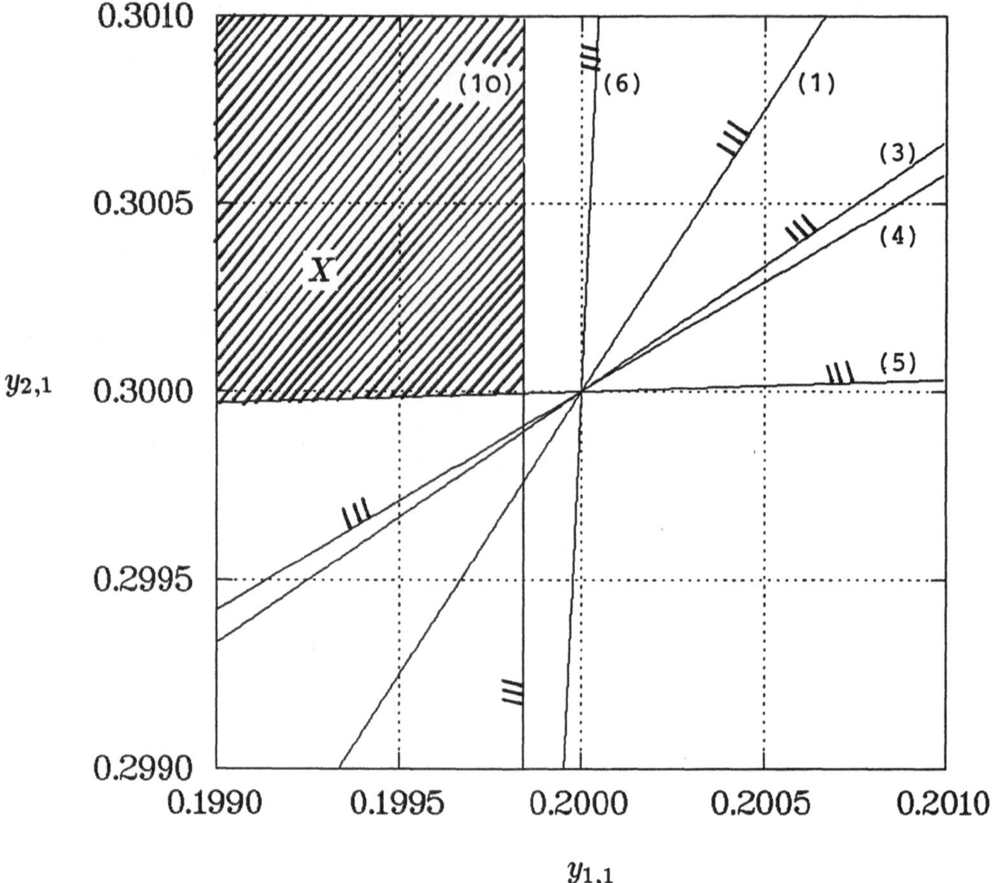

Fig. 5.2.2.2
Enlargement of the environment of point $(y_{1,1}, y_{2,1}) = (0,2;0,3)$ in Fig. 5.2.2.1

Tab. 5.2.2.1
Tableaux of the cycling example (5.2.2.4)
with simplex cycle C in (5.2.1.2)

	1	2	3	4	5	6
5	$\frac{1}{10}$	-60	$-\frac{1}{25}$	9	1	0
6	$\frac{7}{10}$	-90	$-\frac{1}{50}$	3	0	1
Δz_j	$-\frac{1}{2}$	150	$-\frac{1}{50}$	6	0	0
1	1	-600	$-\frac{2}{5}$	90	10	0
6	0	330	$\frac{13}{50}$	-60	-7	1
Δz_j	0	-150	$-\frac{11}{50}$	51	5	0
1	1	0	$\frac{4}{55}$	$-\frac{210}{11}$	$-\frac{30}{11}$	$\frac{20}{11}$
2	0	1	$\frac{13}{16500}$	$-\frac{2}{11}$	$-\frac{7}{330}$	$\frac{1}{330}$
Δz_j	0	0	$-\frac{28}{275}$	$\frac{261}{11}$	$\frac{20}{11}$	$\frac{5}{11}$
3	$\frac{55}{4}$	0	1	$-\frac{525}{2}$	$-\frac{75}{2}$	25
2	$-\frac{13}{1200}$	1	0	$\frac{1}{40}$	$\frac{1}{120}$	$-\frac{1}{60}$
Δz_j	$\frac{7}{5}$	0	0	-3	-2	3
3	-100	10500	1	0	50	-150
4	$-\frac{13}{30}$	40	0	1	$\frac{1}{3}$	$-\frac{2}{3}$
Δz_j	$\frac{1}{10}$	120	0	0	-1	1
5	-2	210	$\frac{1}{50}$	0	1	-3
4	$\frac{7}{30}$	-30	$-\frac{1}{150}$	1	0	$\frac{1}{3}$
Δz_j	$-\frac{19}{10}$	330	$\frac{1}{50}$	0	0	-2

The next pivot step generates the initial tableau.

Legend: Tableau associated with basis B

	column indices
indices of the basis B	$B^{-1}\bar{Y}$
Δz_j	$c_B^T B^{-1}\bar{Y} - \bar{d}$

(The right-hand side always equals zero and is therefore omitted.)

5.2.3 ADDITION OF A COLUMN TO THE INITIAL TABLEAU

The above construction steps (row or column modifications) do not change the dimensions of the cycling example (row and column numbers of the tableaux). In the following we construct an *additional* column to the initial tableau (cf. also Section 5.2.4), which occurs at least once as a basic column of the cycling example constructed in this way.[155]

Our starting point is again the cycling example (5.2.1.1) with simplex cycle C in (5.2.1.2).

The task is now the determination of additional coefficients $y_{1,7}$, $y_{2,7}$, d_7 in (5.2.1.1), such that

$$\left.\begin{array}{llllllll}
\max z = & \frac{3}{4}x_1 & -150x_2 & +\frac{1}{50}x_3 & -6x_4 & & & +d_7x_7 & \\
\text{s.t.} & & & & & & & & \\
& \frac{1}{4}x_1 & -60x_2 & -\frac{1}{25}x_3 & +9x_4 & +x_5 & & +y_{1,7}x_7 & =0 \\
& \frac{1}{2}x_1 & -90x_2 & -\frac{1}{50}x_3 & +3x_4 & & +x_6 & +y_{2,7}x_7 & =0 \\
& & & & & & & x_1,\ldots,x_7 & \geq 0
\end{array}\right\} \tag{5.2.3.1}$$

is a cycling example with any "enlarged" simplex cycle[156], which can e.g. have the form

$$C' = (\{1,2\},\{2,3\},\{3,4\},\{4,5\},\{5,6\},\{1,6\},\{1,7\},\{1,2\}). \tag{5.2.3.2}$$

According to Theorem 4.4.2.1 (cf. also Sections 5.2.1 and 5.2.2), problem (5.2.3.1) is a cycling example with simplex cycle C' if and only if the corresponding system

---

[155] This means that the "new" column must be chosen as the pivot column in any iteration (cf. (5.2.3.7) and Tab. 5.2.3.1).

[156] The enlarged simplex cycle must contain a set with index 7.

$$\left.\begin{array}{l}
D_{1,2}, D_{3,2}, D_{3,4}, D_{5,4}, D_{1,6}, D_{1,7} > 0 \\
\bar{D}_{1,2,3} \begin{cases} < 0 \\ \leq \bar{D}_{1,2,4}, \bar{D}_{1,2,5}, \bar{D}_{1,2,6}, \bar{D}_{1,2,7} \end{cases} \\
\bar{D}_{3,2,4} \begin{cases} < 0 \\ \leq \bar{D}_{3,2,5}, \bar{D}_{3,2,6}, \bar{D}_{3,2,1}, \bar{D}_{3,2,7} \end{cases} \\
\bar{D}_{3,4,5} \begin{cases} < 0 \\ \leq \bar{D}_{3,4,6}, \bar{D}_{3,4,1}, \bar{D}_{3,4,2}, \bar{D}_{3,4,7} \end{cases} \\
\bar{D}_{5,4,6} \begin{cases} < 0 \\ \leq \bar{D}_{5,4,1}, \bar{D}_{5,4,2}, \bar{D}_{5,4,3}, \bar{D}_{5,4,7} \end{cases} \\
\bar{D}_{5,6,1} \begin{cases} < 0 \\ \leq \bar{D}_{5,6,2}, \bar{D}_{5,6,3}, \bar{D}_{5,6,4}, \bar{D}_{5,6,7} \end{cases} \\
\bar{D}_{1,6,7} \begin{cases} < 0 \\ \leq \bar{D}_{1,6,3}, \bar{D}_{1,6,4}, \bar{D}_{1,6,5}, \bar{D}_{1,6,2} \end{cases} \\
\bar{D}_{1,7,2} \begin{cases} < 0 \\ \leq \bar{D}_{1,7,3}, \bar{D}_{1,7,4}, \bar{D}_{1,7,5}, \bar{D}_{1,7,6} \end{cases}
\end{array}\right\} \qquad (5.2.3.3)$$

is satisfied. An inequality of the above system can only contain any one of the variables $y_{1,7}, y_{2,7}, d_7$ if the index 7 occurs in at least one of its determinants. Thus we can simplify (5.2.3.3) as follows (cf. the footnote to (5.2.1.5)):

$$\left.\begin{array}{l}
D_{1,7} > 0 \\
\bar{D}_{1,2,3} \leq \bar{D}_{1,2,7} \\
\bar{D}_{3,2,4} \leq \bar{D}_{3,2,7} \\
\bar{D}_{3,4,5} \leq \bar{D}_{3,4,7} \\
\bar{D}_{5,4,6} \leq \bar{D}_{5,4,7} \\
\bar{D}_{5,6,1} \leq \bar{D}_{5,6,7} \\
\bar{D}_{1,6,7} \leq \bar{D}_{1,6,2} \\
\bar{D}_{1,7,2} \begin{cases} < 0 \\ \leq \bar{D}_{1,7,3}, \bar{D}_{1,7,4}, \bar{D}_{1,7,5}, \bar{D}_{1,7,6} \end{cases}
\end{array}\right\} \qquad (5.2.3.4)$$

Inserting the coefficients of (5.2.3.1) in (5.2.3.4) yields the system

$$\left.\begin{array}{r}
-\frac{1}{2}y_{1,7}+\frac{1}{4}y_{2,7}>0 \\
-\frac{3}{5}\le\frac{15}{2}y_{1,7}+\frac{15}{2}y_{2,7}-\frac{15}{2}d_7 \\
-\frac{36}{5}\le-\frac{24}{5}y_{1,7}+\frac{36}{5}y_{2,7}-\frac{12}{5}d_7 \\
-\frac{3}{50}\le-\frac{3}{50}y_{1,7}+\frac{3}{50}y_{2,7}-\frac{3}{50}d_7 \\
-6\le-6y_{2,7}-3d_7 \\
-\frac{3}{4}\le-d_7 \\
\frac{3}{4}y_{1,7}-\frac{1}{4}d_7\le-\frac{15}{2} \\
-\frac{15}{2}y_{1,7}-\frac{15}{2}y_{2,7}+\frac{15}{2}d_7\left\{\begin{array}{l}
<0 \\
\le\frac{1}{40}y_{1,7}-\frac{7}{200}y_{2,7}+\frac{3}{200}d_7 \\
\le-\frac{21}{4}y_{1,7}+\frac{33}{4}y_{2,7}-\frac{15}{4}d_7 \\
\le\frac{3}{4}y_{2,7}-\frac{1}{2}d_7 \\
\le-\frac{3}{4}y_{1,7}+\frac{1}{4}d_7
\end{array}\right.
\end{array}\right\}\tag{5.2.3.5}$$

Transforming (5.2.3.5) into the “normal form” (cf. Sections 5.2.1 and 5.2.2) yields:

$$\left.\begin{array}{rrrll}
2y_{1,7}\;- & y_{2,7} & & <0 & (1) \\
-y_{1,7}\;- & y_{2,7}\;+ & d_7 & \le\frac{2}{25} & (2) \\
2y_{1,7}\;- & 3y_{2,7}\;+ & d_7 & \le 3 & (3) \\
y_{1,7}\;- & y_{2,7}\;+ & d_7 & \le 1 & (4) \\
 & 2y_{2,7}\;+ & d_7 & \le 2 & (5) \\
 & & d_7 & \le\frac{3}{4} & (6) \\
3y_{1,7}\;- & & d_7 & \le -30 & (7) \\
-y_{1,7}\;- & y_{2,7}\;+ & d_7 & <0 & (8) \\
-1505y_{1,7}\;- & 1493y_{2,7}\;+ & 1497d_7 & \le 0 & (9) \\
-9y_{1,7}\;- & 63y_{2,7}\;+ & 45d_7 & \le 0 & (10) \\
-30y_{1,7}\;- & 33y_{2,7}\;+ & 32d_7 & \le 0 & (11) \\
-27y_{1,7}\;- & 30y_{2,7}\;+ & 29d_7 & \le 0 & (12)
\end{array}\right\}\tag{5.2.3.6}$$

Hence (5.2.3.1) is a cycling example with simplex cycle C' in (5.2.3.2)

Tab. 5.2.3.1
Tableaux of the cycling example (5.2.3.7)
with simplex cycle C' in (5.2.3.2)

	1	2	3	4	5	6	7
5	$\frac{1}{4}$	-60	$-\frac{1}{25}$	9	1	0	-120
6	$\frac{1}{2}$	-90	$-\frac{1}{50}$	3	0	1	-180
Δz_j	$-\frac{3}{4}$	150	$-\frac{1}{50}$	6	0	0	301
1	1	-240	$-\frac{4}{25}$	36	4	0	-480
6	0	30	$\frac{3}{50}$	-15	-2	1	60
Δz_j	0	-30	$-\frac{7}{50}$	33	3	0	-59
1	1	0	$\frac{8}{25}$	-84	-12	8	0
7	0	$\frac{1}{2}$	$\frac{1}{1000}$	$-\frac{1}{4}$	$-\frac{1}{30}$	$\frac{1}{60}$	1
Δz_j	0	$-\frac{1}{2}$	$-\frac{81}{1000}$	$\frac{73}{4}$	$\frac{31}{30}$	$\frac{59}{60}$	0
1	1	0	$\frac{8}{25}$	-84	-12	8	0
2	0	1	$\frac{1}{500}$	$-\frac{1}{2}$	$-\frac{1}{15}$	$\frac{1}{30}$	2
Δz_j	0	0	$-\frac{2}{25}$	18	1	1	1
3	$\frac{25}{8}$	0	1	$-\frac{525}{2}$	$-\frac{75}{2}$	25	0
2	$-\frac{1}{160}$	1	0	$\frac{1}{40}$	$\frac{1}{120}$	$-\frac{1}{60}$	2
Δz_j	$\frac{1}{4}$	0	0	-3	-2	3	1
3	$-\frac{125}{2}$	10500	1	0	50	-150	21000
4	$-\frac{1}{4}$	40	0	1	$\frac{1}{3}$	$-\frac{2}{3}$	80
Δz_j	$-\frac{1}{2}$	120	0	0	-1	1	241
5	$-\frac{5}{4}$	210	$\frac{1}{50}$	0	1	-3	420
4	$\frac{1}{6}$	-30	$-\frac{1}{150}$	1	0	$\frac{1}{3}$	-60
Δz_j	$-\frac{7}{4}$	330	$\frac{1}{50}$	0	0	-2	661

The next pivot step generates the initial tableau.

Legend: Tableau associated with basis B

	column indices
indices of the basis B	$B^{-1}\bar{Y}$
Δz_j	$c_B^T B^{-1}\bar{Y} - \bar{d}$

(The right-hand side always equals zero and is therefore omitted.)

if and only if the coefficients $y_{1,7}, y_{2,7}, d_7$ in (5.2.3.1) satisfy the above system. A solution of (5.2.3.6) is

$$(y_{1,7}, y_{2,7}, d_7) = (-120, -180, -301).$$

Inserting these values in (5.2.3.1) leads to the cycling example

$$\left.\begin{array}{llllllll} \max z = & \frac{3}{4}x_1 & -150x_2 & +\frac{1}{50}x_3 & -6x_4 & & & -301x_7 & \\ \text{s.t.} & & & & & & & & \\ & \frac{1}{4}x_1 & -60x_2 & +\frac{1}{25}x_3 & +9x_4 & +x_5 & & -120x_7 & = 0 \\ & \frac{1}{2}x_1 & -90x_2 & -\frac{1}{50}x_3 & +3x_4 & & +x_6 & -180x_7 & = 0 \\ & & & & & & & x_1,\dots,x_7 & \geq 0 \end{array}\right\} \quad (5.2.3.7)$$

with simplex cycle C' in (5.2.3.2).

The associated tableaux are listed in Tab. 5.2.3.1.

5.2.4 ADDITION OF A ROW TO THE INITIAL TABLEAU

In order to increase the row dimension as well (cf. Section 5.2.3), we add another row and another column in a single operation to the initial tableau. The additional row and column are unit vectors of appropriate dimensions. The facts will be illustrated again by means of the cycling example (5.2.1.1) with simplex cycle C in (5.2.1.2).

From this we obtain the cycling example

$$\left.\begin{array}{lllllllll} \max z = & \frac{3}{4}x_1 & - 150x_2 & + \frac{1}{50}x_3 & - 6x_4 & & & & \\ \text{s.t.} & & & & & & & & \\ & \frac{1}{4}x_1 & - 60x_2 & - \frac{1}{25}x_3 & + 9x_4 & + x_5 & & & = 0 \\ & \frac{1}{2}x_1 & - 90x_2 & - \frac{1}{50}x_3 & + 3x_4 & & +x_6 & & = 0 \\ & & & & & & & x_7 & = 0 \\ & & & & & & & x_1,\dots,x_7 & \geq 0 \end{array}\right\} \quad (5.2.4.1)$$

with simplex cycle

$$C'' = (\{1,2,7\},\{2,3,7\},\{3,4,7\},\{4,5,7\},\{5,6,7\},\{1,6,7\},\{1,2,7\}) \quad (5.2.4.2)$$

On grounds of its construction, (5.2.4.1) is "reducible" to a cycling example with two constraints, in so far as the index sets of C'' in (5.2.4.2) can be biuniquely assigned to the index sets of C in (5.2.1.2). However, we obtain a nonreducible cycling example from (5.2.4.1) by constructing an additional column as in Section 5.2.3. In so doing we must determine additional coefficients $y_{1.8}, y_{2.8}, y_{3.8}, d_8$ in (5.2.4.1) such that

$$\left.\begin{array}{llllllll}
\max z = \frac{3}{4}x_1 & -150x_2 & +\frac{1}{50}x_3 & -6x_4 & & & +d_8x_8 & \\
\text{s.t.} & & & & & & & \\
\frac{1}{4}x_1 & -60x_2 & -\frac{1}{25}x_3 & +9x_4+x_5 & & & +y_{1,8}x_8 & =0 \\
\frac{1}{2}x_1 & -90x_2 & -\frac{1}{50}x_3 & +3x_4 & +x_6 & & +y_{2,8}x_8 & =0 \\
 & & & & & x_7 & +y_{3,8}x_8 & =0 \\
 & & & & & & x_1,\dots,x_8 & \geq 0
\end{array}\right\} \tag{5.2.4.3}$$

is a cycling example with an "enlarged" simplex cycle, e.g. with

$$C'' = (\{1,2,7\},\{3,2,7\},\{3,4,7\},\{5,4,7\},\{5,6,7\},\{1,6,7\},\{1,6,8\}, \{1,2,8\},\{1,2,7\}) \tag{5.2.4.4}$$

(cf. also (5.2.3.2)). To this end we solve the corresponding determinant inequality system (cf. Theorem 4.4.2.1):

$$\left.\begin{array}{l}
D_{1,2,7}, D_{3,2,7}, D_{3,4,7}, D_{5,4,7}, D_{5,6,7}, D_{1,6,7}, D_{1,6,8}, D_{1,2,8} > 0 \\
\bar{D}_{1,2,7,3} \begin{cases} < 0 \\ \leq \bar{D}_{1,2,7,4}, \bar{D}_{1,2,7,5}, \bar{D}_{1,2,7,6}, \bar{D}_{1,2,7,8} \end{cases} \\
\bar{D}_{3,2,7,4} \begin{cases} < 0 \\ \leq \bar{D}_{3,2,7,5}, \bar{D}_{3,2,7,6}, \bar{D}_{3,2,7,1}, \bar{D}_{3,2,7,8} \end{cases} \\
\bar{D}_{3,4,7,5} \begin{cases} < 0 \\ \leq \bar{D}_{3,4,7,6}, \bar{D}_{3,4,7,1}, \bar{D}_{3,4,7,2}, \bar{D}_{3,4,7,8} \end{cases} \\
\bar{D}_{5,4,7,6} \begin{cases} < 0 \\ \leq \bar{D}_{5,4,7,1}, \bar{D}_{5,4,7,2}, \bar{D}_{5,4,7,3}, \bar{D}_{5,4,7,8} \end{cases} \\
\bar{D}_{5,6,7,1} \begin{cases} < 0 \\ \leq \bar{D}_{5,6,7,2}, \bar{D}_{5,6,7,3}, \bar{D}_{5,6,7,4}, \bar{D}_{5,6,7,8} \end{cases} \\
\bar{D}_{1,6,7,8} \begin{cases} < 0 \\ \leq \bar{D}_{1,6,7,2}, \bar{D}_{1,6,7,3}, \bar{D}_{1,6,7,4}, \bar{D}_{1,6,7,5} \end{cases} \\
\bar{D}_{1,6,8,2} \begin{cases} < 0 \\ \leq \bar{D}_{1,6,8,3}, \bar{D}_{1,6,8,4}, \bar{D}_{1,6,8,5}, \bar{D}_{1,6,8,7} \end{cases} \\
\bar{D}_{1,2,8,7} \begin{cases} < 0 \\ \leq \bar{D}_{1,2,8,3}, \bar{D}_{1,2,8,4}, \bar{D}_{1,2,8,5}, \bar{D}_{1,2,8,6} \end{cases}
\end{array}\right\} \tag{5.2.4.5}$$

From this we obtain by similar transformations as in Section 5.2.3 (cf. (5.2.3.3)-(5.2.3.6)):

$$\left.\begin{array}{rrrrll}
 & & -y_{3,8} & & < 0 & (1) \\
2y_{1,8} - & 3y_{2,8} + & & d_8 & \leq 3 & (2) \\
 & 2y_{2,8} + & & d_8 & \leq 2 & (3) \\
 & & & d_8 & \leq \frac{3}{4} & (4) \\
3y_{1,8} - & & & d_8 & \leq -30 & (5) \\
3y_{1,8} - & & 30y_{3,8} - & d_8 & \leq 0 & (6) \\
-y_{1,8} - & y_{2,8} + & \frac{2}{25}y_{3,8} + & d_8 & \leq 0 & (7)
\end{array}\right\} \tag{5.2.4.6}$$

A solution of (5.2.4.6) is $(y_{1,8}, y_{2,8}, y_{3,8}, d_8) = (-20, 1, 1, -21)$. This leads to the nonreducible cycling example

Tab. 5.2.4.1
Tableaux of the cycling example (5.2.4.7)
with simplex cycle C'' in (5.2.4.4)

	1	2	3	4	5	6	7	8
5	$\frac{1}{4}$	-60	$-\frac{1}{25}$	9	1	0	0	-20
6	$\frac{1}{2}$	-90	$-\frac{1}{50}$	3	0	1	0	1
7	0	0	0	0	0	0	1	1
Δz_j	$-\frac{3}{4}$	150	$-\frac{1}{50}$	6	0	0	0	21
1	1	-240	$-\frac{4}{25}$	36	4	0	0	-80
6	0	30	$\frac{3}{50}$	-15	-2	1	0	41
7	0	0	0	0	0	0	1	1
Δz_j	0	-30	$-\frac{7}{50}$	33	3	0	0	-39
1	1	-240	$-\frac{4}{25}$	36	4	0	80	0
6	0	30	$\frac{3}{50}$	-15	-2	1	-41	0
8	0	0	0	0	0	0	1	1
Δz_j	0	-30	$-\frac{7}{50}$	33	3	0	39	0
1	1	0	$\frac{8}{25}$	-84	-12	8	-248	0
2	0	1	$\frac{1}{500}$	$-\frac{1}{2}$	$-\frac{1}{15}$	$\frac{1}{30}$	$-\frac{41}{30}$	0
8	0	0	0	0	0	0	1	1
Δz_j	0	0	$-\frac{2}{25}$	18	1	1	-2	0
1	1	0	$\frac{8}{25}$	-84	-12	8	0	248
2	0	1	$\frac{1}{500}$	$-\frac{1}{2}$	$-\frac{1}{15}$	$\frac{1}{30}$	0	$\frac{41}{30}$
7	0	0	0	0	0	0	1	1
Δz_j	0	0	$-\frac{2}{25}$	18	1	1	0	2

Tab. 5.2.4.1 (continued)

3	$\frac{25}{8}$	0	1	$-\frac{525}{2}$	$-\frac{75}{2}$	25	0	775
2	$-\frac{1}{160}$	1	0	$\frac{1}{40}$ (circled)	$\frac{1}{120}$	$-\frac{1}{60}$	0	$-\frac{11}{60}$
7	0	0	0	0	0	0	1	1
Δz_j	$\frac{1}{4}$	0	0	-3	-2	3	0	64
3	$-\frac{125}{2}$	10500	1	0	50 (circled)	-150	0	-1150
4	$-\frac{1}{4}$	40	0	1	$\frac{1}{3}$	$-\frac{2}{3}$	0	$-\frac{22}{3}$
7	0	0	0	0	0	0	1	1
Δz_j	$-\frac{1}{2}$	120	0	0	-1	1	0	42
5	$-\frac{5}{4}$	210	$\frac{1}{50}$	0	1	-3	0	-23
4	$\frac{1}{6}$	-30	$-\frac{1}{150}$	1	0	$\frac{1}{3}$ (circled)	0	$\frac{1}{3}$
7	0	0	0	0	0	0	1	1
Δz_j	$-\frac{7}{4}$	330	$\frac{1}{50}$	0	0	-2	0	19

The next pivot step generates the initial tableau.

Legend: Tableau associated with basis B

	column indices
indices of the basis B	$B^{-1}\bar{Y}$
Δz_j	$c_B^T B^{-1}\bar{Y} - \bar{d}$

(The right-hand side always equals zero and is therefore omitted.)

$$\left.\begin{array}{lllllllll} \max z = & \frac{3}{4}x_1 & -150x_2 & +\frac{1}{50}x_3 & -6x_4 & & & -21x_8 & \\ \text{s.t.} & & & & & & & & \\ & \frac{1}{4}x_1 & -60x_2 & -\frac{1}{25}x_3 & +9x_4 & +x_5 & & -20x_8 & =0 \\ & \frac{1}{2}x_1 & -90x_2 & -\frac{1}{50}x_3 & +3x_4 & & +x_6 & +x_8 & =0 \\ & & & & & & & x_7+x_8 & =0 \\ & & & & & & & x_1,\dots,x_8 & \geq 0 \end{array}\right\} \quad (5.2.4.7)$$

with simplex cycle C'' in (5.2.4.4). The associated tableaux are listed in Tab. 5.2.4.1.

5.2.5 COMBINATION OF CONSTRUCTION STEPS

There are various possibilities for combining the above construction steps. In the following we present algorithmic descriptions for two examples of combined procedures. The algorithms are based on the Theorems 5.2.5.1.1 and 5.2.5.1.2 below, which enable us performing construction steps for *arbitrary* elements in the initial tableau of the given cycling example.[157]

For the sake of clarity we take a cycling example with two constraints and four structural variables as a basis (cf. (5.2.1) and (5.2.2)):

$$\left.\begin{array}{lllll} \max z = & d_1x_1+ & \dots & +d_6x_6 & \\ \text{s.t.} & & & & \\ & y_{1,1}x_1+ & \dots & +y_{1,6}x_6 & =0 \\ & y_{2,1}x_1+ & \dots & +y_{2,6}x_6 & =0 \\ & & & x_1,\dots,x_6 & \geq 0 \end{array}\right\} \qquad (5.2.5.1)$$

where

$$\begin{pmatrix} y_{1,5}, & y_{1,6} \\ y_{2,5}, & y_{2,6} \\ d_5, & d_6 \end{pmatrix} = \begin{pmatrix} 1, & 0 \\ 0, & 1 \\ 0, & 0 \end{pmatrix}.$$

---

157 In Sections 5.2.1 – 5.2.4 we confined ourselves to illustrating the construction steps by means of tableaux with specific numerical values.

Let

$$C = (I_1, \ldots, I_k, I_1) \tag{5.2.5.2}$$

be the simplex cycle of (5.2.5.1) and let

$$T = \begin{pmatrix} y_{1,1}, & y_{1,2}, & y_{1,3}, & y_{1,4}, & y_{1,5}, & y_{1,6} \\ y_{2,1}, & y_{2,2}, & y_{2,3}, & y_{2,4}, & y_{2,5}, & y_{2,6} \\ -d_1, & -d_2, & -d_3, & -d_4, & -d_5, & -d_6 \end{pmatrix} \tag{5.2.5.3}$$

denote the initial tableau of (5.2.5.1).

5.2.5.1 SUCCESSIVE MODIFICATION OF ROWS

The procedure below starts with the tableau T of a cycling example (5.2.5.1). All rows of T are modified successively (cf. Section 5.2.1). The tableau T_1 resulting from these transformations is the initial tableau of a new cycling example of the form (5.2.5.1) (with the simplex cycle C in (5.2.5.2)) and is printed.

Subsequently, by multiple row modifications, T_1 is transformed into a tableau T_2 which is printed also, etc. The whole process continues until the required number of tableaux is generated.

In order to permit a unique description for all row modifications (cf. Theorem 5.2.5.1.1), we arrange the notation

$$y_{3,j} = -d_j \quad \text{for} \quad j = 1, \ldots, 6 \tag{5.2.5.1.1}$$

for the present section, i.e. the tableau T in (5.2.5.3) has the form

$$T = \begin{pmatrix} y_{1,1}, \ldots, y_{1,6} \\ y_{2,1}, \ldots, y_{2,6} \\ y_{3,1}, \ldots, y_{3,6} \end{pmatrix} = \begin{pmatrix} y_{1,1}, \ldots, y_{1,4}, 1, 0 \\ y_{2,1}, \ldots, y_{2,4}, 0, 1 \\ y_{3,1}, \ldots, y_{3,4}, 0, 0 \end{pmatrix} \tag{5.2.5.1.2}$$

Beside the definitions in Section 4.4.1, we make use of the following notation for determinants of T:

Let $\bar{D}^i_{j,k}$ denote the determinant of the matrix resulting from deletion of the i-th row in

$$\begin{pmatrix} y_{1,j}, & y_{1,k} \\ y_{2,j}, & y_{2,k} \\ y_{3,j}, & y_{3,k} \end{pmatrix}$$

$$(j, k \in \{1, \ldots, 6\}, i \in \{1, \ldots, 3\}).$$

For example it holds

$$\bar{D}^2_{5,4} = \begin{vmatrix} y_{1,5}, & y_{1,4} \\ y_{3,5}, & y_{3,4} \end{vmatrix} = y_{1,5}y_{3,4} - y_{3,5}y_{1,4}.$$

Moreover we define

$$D^i_j = \begin{cases} y_{2,j} & \text{for } i = 1 \\ y_{1,j} & \text{for } i = 2 \end{cases}$$

$$(j \in \{1, \ldots, 6\}, i \in \{1, 2\}).$$

For example it holds $D^2_6 = y_{1,6}$.[158]

The algorithm consists of the following steps, where step 1 will be subdivided later:

Input data:

Initial tableau $T = (y_{i,j})$
in (5.2.5.1.2) ($i = 1, 2, 3$; $j = 1, \ldots, 6$),
number p of required cycling examples,
tableau counter $s = 1$,
row counter $i = 0$.

Step 1 (Row modification):

Set $i = i + 1$,
determine a modified i-th row
$(\tilde{y}_{i,1}, \ldots, \tilde{y}_{i,4})$ in T as in Section 5.2.1
($y_{i,5}, y_{i,6}$ remain unchanged),[159]
set $y_{i,j} = \tilde{y}_{i,j}$ for $j = 1, \ldots, 4$.

---

[158] We can interpret D^i_j as the determinant of a matrix, consisting of one element.

[159] Cf. Theorem 5.2.5.1.1.

Step 2 (Printing of a modified initial tableau):
If $i < 3$, go to step 1,
else print tableau $T = (y_{i,j})$
(i = 1,2,3; j = 1,...,6).

Step 3 (Test of stopping criterion):
If $s < p$, set $i = 0$, $s = s + 1$ and go to step 1;
else: STOP – the required number of tableaux has been printed.

The following statement permits a subdivision of step 1 (row modification):[160]

Theorem 5.2.5.1.1: Let the cycling example (5.2.5.1) with simplex cycle C in (5.2.5.2) and the initial tableau T in (5.2.5.1.2) be given. For any fixed index $i \in \{1, 2, 3\}$ let the elements $y_{i,1}, \ldots, y_{i,4}$ in T be variable, while the other elements of T are fixed.

Then tableau T is an initial tableau of a cycling example with simplex cycle C if and only if the following system of linear inequalities is satisfied (the inequalities (1) – (5) can be ignored in the case $i = 3$):[161]

[160] Cf. the footnote to the introductory representations in Section 5.2.5.

[161] The coefficients and the right-hand side in (5.2.5.1.3) are fixed, since all elements of T except $y_{i,1},\ldots,y_{i,4}$ are assumed to be fixed.

$$
(-1)^{i+1}\left(\begin{array}{|c|c|c|c|}
-D^i_2 & D^i_1 & 0 & 0 \\
0 & D^i_3 & -D^i_2 & 0 \\
0 & 0 & -D^i_4 & D^i_3 \\
0 & 0 & 0 & D^i_5 \\
-D^i_6 & 0 & 0 & 0 \\ \hline
\bar{D}^i_{2,3} & -\bar{D}^i_{1,3} & \bar{D}^i_{1,2} & 0 \\
\bar{D}^i_{2,3}-\bar{D}^i_{2,4} & \bar{D}^i_{1,4}-\bar{D}^i_{1,3} & \bar{D}^i_{1,2} & -\bar{D}^i_{1,2} \\
\bar{D}^i_{2,3}-\bar{D}^i_{2,5} & \bar{D}^i_{1,5}-\bar{D}^i_{1,3} & \bar{D}^i_{1,2} & 0 \\
\bar{D}^i_{2,3}-\bar{D}^i_{2,6} & \bar{D}^i_{1,6}-\bar{D}^i_{1,3} & \bar{D}^i_{1,2} & 0 \\ \hline
0 & -\bar{D}^i_{3,4} & \bar{D}^i_{2,4} & \bar{D}^i_{3,2} \\
0 & \bar{D}^i_{3,5}-\bar{D}^i_{3,4} & \bar{D}^i_{2,4}-\bar{D}^i_{2,5} & \bar{D}^i_{3,2} \\
0 & \bar{D}^i_{3,6}-\bar{D}^i_{3,4} & \bar{D}^i_{2,4}-\bar{D}^i_{2,6} & \bar{D}^i_{3,2} \\
-\bar{D}^i_{3,2} & \bar{D}^i_{3,1}-\bar{D}^i_{3,4} & \bar{D}^i_{2,4}-\bar{D}^i_{2,1} & \bar{D}^i_{3,2} \\ \hline
0 & 0 & \bar{D}^i_{4,5} & -\bar{D}^i_{3,5} \\
0 & 0 & \bar{D}^i_{4,5}-\bar{D}^i_{4,6} & \bar{D}^i_{3,6}-\bar{D}^i_{3,5} \\
-\bar{D}^i_{3,4} & 0 & \bar{D}^i_{4,5}-\bar{D}^i_{4,1} & \bar{D}^i_{3,1}-\bar{D}^i_{3,5} \\
0 & -\bar{D}^i_{3,4} & \bar{D}^i_{4,5}-\bar{D}^i_{4,2} & \bar{D}^i_{3,2}-\bar{D}^i_{3,5} \\ \hline
0 & 0 & 0 & -\bar{D}^i_{5,6} \\
-\bar{D}^i_{5,4} & 0 & 0 & \bar{D}^i_{5,1}-\bar{D}^i_{5,6} \\
0 & -\bar{D}^i_{5,4} & 0 & \bar{D}^i_{5,2}-\bar{D}^i_{5,6} \\
0 & 0 & -\bar{D}^i_{5,4} & \bar{D}^i_{5,3}-\bar{D}^i_{5,6} \\ \hline
\bar{D}^i_{5,6} & 0 & 0 & 0 \\
\bar{D}^i_{5,6} & -\bar{D}^i_{5,6} & 0 & 0 \\
\bar{D}^i_{5,6} & 0 & -\bar{D}^i_{5,6} & 0 \\
\bar{D}^i_{5,6} & 0 & 0 & -\bar{D}^i_{5,6} \\ \hline
\bar{D}^i_{6,2} & \bar{D}^i_{1,6} & 0 & 0 \\
\bar{D}^i_{6,2}-\bar{D}^i_{6,3} & \bar{D}^i_{1,6} & -\bar{D}^i_{1,6} & 0 \\
\bar{D}^i_{6,2}-\bar{D}^i_{6,4} & \bar{D}^i_{1,6} & 0 & -\bar{D}^i_{1,6} \\
\bar{D}^i_{6,2}-\bar{D}^i_{6,5} & \bar{D}^i_{1,6} & 0 & 0 \\ \hline
\end{array}\right)
\begin{pmatrix} y_{i,1} \\ y_{i,2} \\ y_{i,3} \\ y_{i,4} \end{pmatrix}
\left\{\begin{array}{c}
< \\ < \\ < \\ < \\ < \\ \hline
< \\ \leq \\ \leq \\ \leq \\ \hline
< \\ \leq \\ \leq \\ \leq \\ \hline
< \\ \leq \\ \leq \\ \leq \\ \hline
< \\ \leq \\ \leq \\ \leq \\ \hline
< \\ \leq \\ \leq \\ \leq \\ \hline
< \\ \leq \\ \leq \\ \leq \\ \hline
\end{array}\right\}
(-1)^{i+1}\left(\begin{array}{cc}
0 & \\
0 & \\
0 & \\
D^i_4 y_{i,5} & \\
 & -D^i_1 y_{i,6} \\ \hline
0 & \\
0 & \\
\bar{D}^i_{1,2} y_{i,5} & \\
 & \bar{D}^i_{1,2} y_{i,6} \\ \hline
0 & \\
\bar{D}^i_{3,2} y_{i,5} & \\
 & \bar{D}^i_{3,2} y_{i,6} \\
0 & \\ \hline
-\bar{D}^i_{3,4} y_{i,5} & \\
-\bar{D}^i_{3,4} y_{i,5} & +\bar{D}^i_{3,4} y_{i,6} \\
-\bar{D}^i_{3,4} y_{i,5} & \\
-\bar{D}^i_{3,4} y_{i,5} & \\ \hline
-\bar{D}^i_{4,6} y_{i,5} & -\bar{D}^i_{5,4} y_{i,6} \\
(\bar{D}^i_{4,1}-\bar{D}^i_{4,6}) y_{i,5} & -\bar{D}^i_{5,4} y_{i,6} \\
(\bar{D}^i_{4,2}-\bar{D}^i_{4,6}) y_{i,5} & -\bar{D}^i_{5,4} y_{i,6} \\
(\bar{D}^i_{4,3}-\bar{D}^i_{4,6}) y_{i,5} & -\bar{D}^i_{5,4} y_{i,6} \\ \hline
-\bar{D}^i_{6,1} y_{i,5} & +\bar{D}^i_{5,1} y_{i,6} \\
(\bar{D}^i_{6,2}-\bar{D}^i_{6,1}) y_{i,5} & +(\bar{D}^i_{5,1}-\bar{D}^i_{5,2}) y_{i,6} \\
(\bar{D}^i_{6,3}-\bar{D}^i_{6,1}) y_{i,5} & +(\bar{D}^i_{5,1}-\bar{D}^i_{5,3}) y_{i,6} \\
(\bar{D}^i_{6,4}-\bar{D}^i_{6,1}) y_{i,5} & +(\bar{D}^i_{5,1}-\bar{D}^i_{5,4}) y_{i,6} \\ \hline
 & \bar{D}^i_{1,2} y_{i,6} \\
 & (\bar{D}^i_{1,2}-\bar{D}^i_{1,3}) y_{i,6} \\
 & (\bar{D}^i_{1,2}-\bar{D}^i_{1,4}) y_{i,6} \\
\bar{D}^i_{1,6} y_{i,5} & +(\bar{D}^i_{1,2}-\bar{D}^i_{1,5}) y_{i,6} \\ \hline
\end{array}\right)
\begin{array}{c}
(1) \\ (2) \\ (3) \\ (4) \\ (5) \\ (6) \\ (7) \\ (8) \\ (9) \\ (10) \\ (11) \\ (12) \\ (13) \\ (14) \\ (15) \\ (16) \\ (17) \\ (18) \\ (19) \\ (20) \\ (21) \\ (22) \\ (23) \\ (24) \\ (25) \\ (26) \\ (27) \\ (28) \\ (29)
\end{array}
\quad (5.2.5.1.3)
$$

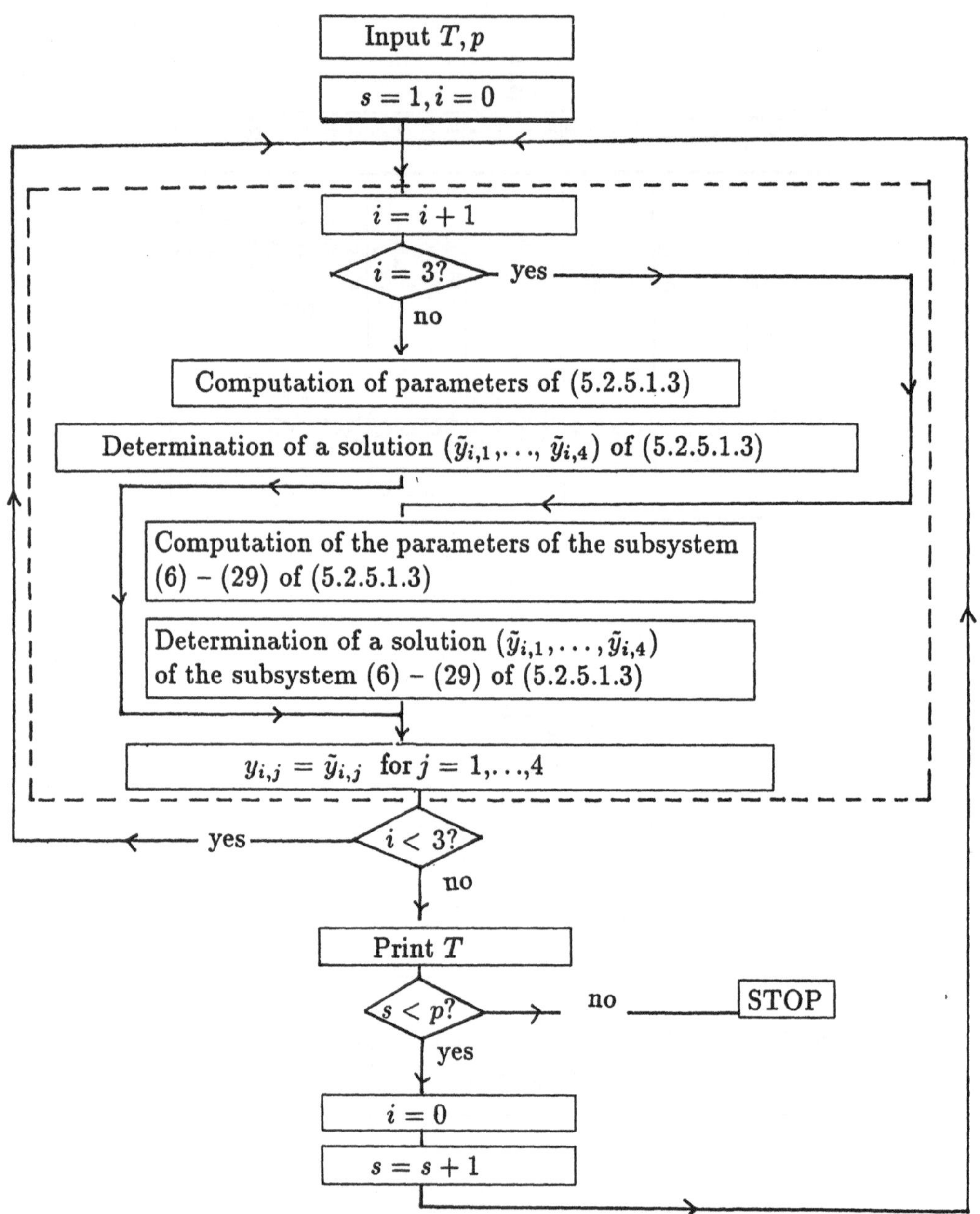

Fig. 5.2.5.1.1
Flow-chart for the successive row modification
(The part in frame corresponds to step 1)

Proof: According to Theorem 4.4.2.1, tableau T is the initial tableau of a cycling example with simplex cycle C if and only if system (5.2.1.4) is satisfied. Expansion of the determinants in (5.2.1.4) by the i-th row ($i \in \{1,2,3\}$) and subsequent arrangement according to the variables $y_{i,1},\ldots,y_{i,4}$ yields the system (5.2.5.1.3).

In the case $i = 3$ the subsystem "$D_{1,2},\ldots,D_{1,6} > 0$" in (5.2.1.4) contains no variables. It is satisfied, since (5.2.5.1) is assumed to be a cycling example. Hence for $i = 3$ the inequalities (1), ..., (5) in (5.2.5.1.3) may be ignored.[162] •

Now we can subdivide step 1 as follows:

1a) Set $i = i + 1$.
1b) If $i = 3$, go to 1c). Else do the following:
Determine the coefficients and the right-hand side of system (5.2.5.1.3) by means of T. Compute a solution $(\tilde{y}_{i,1},\ldots,\tilde{y}_{i,4})$ of (5.2.5.1.3). Go to 1d).
1c) Determine the coefficients and the right-hand side of the subsystem (6) – (29) of (5.2.5.1.3) by means of T. Compute a solution $(\tilde{y}_{i,1},\ldots,\tilde{y}_{i,4})$ of this subsystem.
1d) Set $y_{i,j} = \tilde{y}_{i,j}$ for $j = 1,\ldots,4$.

Fig. 5.2.5.1.1 contains the flow-chart for the above algorithm.

5.2.5.2 SUCCESSIVE ADDITION OF COLUMNS

We start again with the tableau T in (5.2.5.3) belonging to the cycling example (5.2.5.1), where we set

$$\begin{pmatrix} y_{1,1} & y_{1,2} \\ y_{2,1} & y_{2,2} \\ d_1 & d_2 \end{pmatrix} = \begin{pmatrix} 1, & 0 \\ 0, & 1 \\ 0, & 0 \end{pmatrix} \qquad (5.2.5.2.1)$$

instead of the corresponding condition in (5.2.5.1).[163] Let the simplex cycle C in (5.2.5.2) have the special form

[162] Cf. the footnote to system (5.2.1.5).

[163] This is practical in the construction of additional columns.

$$C = (\{1,2\},\{2,3\},\{3,4\},\{4,5\},\{5,6\},\{1,6\},\{1,2\}. \qquad (5.2.5.2.2)$$

By multiple addition of columns to tableau T (cf. Section 5.2.3) we generate the initial tableau of a cycling example of the form

$$\left.\begin{array}{llll} \max z = & d_1x_1+ & \ldots & +d_Nx_N \\ \text{s.t.} & & & \\ & y_{1,1}x_1+ & \ldots & +y_{1,N}x_N & = 0 \\ & y_{2,1}x_1+ & \ldots & +y_{2,N}x_N & = 0 \\ & & & x_1,\ldots,x_N & \geq 0 \end{array}\right\} \qquad (5.2.5.2.3)$$

where $N \geq 7$ denotes a given column number.[164]

The simplex cycle of (5.2.5.2.3) is

$$C^N = (\{1,2\},\{2,3\},\{3,4\},\{4,5\},\{5,6\},\{1,6\},\{1,7\},\ldots,\{1,N\},\{1,2\}) \qquad (5.2.5.2.4)$$

A rough subdivision of the algorithm yields the following steps:

Input data:

Initial tableau

$$T = \begin{pmatrix} y_{1,1} & \ldots & y_{1,6} \\ y_{2,1} & \ldots & y_{2,6} \\ -d_1 & \ldots & -d_6 \end{pmatrix} \quad \text{in (5.2.5.3) (cf. (5.2.5.2.1)),}$$

required column number N in (5.2.5.2.3),

column counter $r = 6$.

---

[164] What the largest possible values for N are must be determined by means of practical tests (cf. Section 5.2.6).

Step 1 (Addition of a column):
Construction of an $(r+1)$-th column $(y_{1,r+1}, y_{2,r+1}, -d_r)^T$ as in Section 5.2.3.[165]

Set $T = \begin{pmatrix} & \mid & y_{1,r+1} \\ T & \mid & y_{2,r+1} \\ & \mid & -d_{r+1} \end{pmatrix}$

Set $r = r + 1$.

Step 2 (Test of stopping criterion):
If $r < N$, go to step 1.
Else: Print tableau T. – STOP.

The following statement serves as a basis for the further subdivision of step 1.[166]

Theorem 5.2.5.2.1: Given a cycling example of the form

$$\left.\begin{array}{lllll} \max z = & d_1x_1+ & \dots & +d_rx_r & \\ \text{s.t.} & & & & \\ & y_{1,1}x_1+ & \dots & +y_{1,r}x_r & = 0 \\ & y_{2,1}x_1+ & \dots & +y_{2,r}x_r & = 0 \\ & & & x_1, \dots, x_r & \geq 0 \end{array}\right\} \quad (5.2.5.2.5)$$

$$\left(\begin{pmatrix} y_{1,1}, & y_{1,2} \\ y_{2,1}, & y_{2,2} \\ d_1, & d_2 \end{pmatrix} = \begin{pmatrix} 1,0 \\ 0,1 \\ 0,0 \end{pmatrix}, \quad r \geq 6 \right)$$

with simplex cycle

$$C^r = (\{1,2\}, \{2,3\}, \{3,4\}, \{4,5\}, \{5,6\}, \{1,6\}, \dots, \{1,r\}, \{1,2\}) \quad (5.2.5.2.6)$$

Then the following holds:

---

[165] Cf. the following Theorem 5.2.5.2.1.

[166] Cf. the footnote to the introductory representations in Section 5.2.5.

The linear optimization problem

$$\left.\begin{array}{llll} \max z = & d_1x_1 + \quad \ldots + & d_{r+1}x_{r+1} & \\ \text{s.t.} & & & \\ & y_{1,1}x_1 + \quad \ldots + & y_{1,r+1}x_{r+1} & = 0 \\ & y_{2,1}x_1 + \quad \ldots + & y_{2,r+1}x_{r+1} & = 0 \\ & & x_1, \ldots, x_{r+1} & \geq 0 \end{array}\right\} \qquad (5.2.5.2.7)$$

($y_{1,j}, y_{2,j}, d_j$ fixed for $j \leq r$; $y_{1,r+1}, y_{2,r+1}, d_{r+1}$ variable) is a cycling example with simplex cycle

$$C^{r+1} = (\{1,2\}, \{2,3\}, \{3,4\}, \{4,5\}, \{5,6\}, \{1,6\}, \ldots, \{1, r+1\}, \{1,2\}), \qquad (5.2.5.2.8)$$

if and only if the following system of linear inequalities is satisfied:[167]

$$\left.\begin{array}{l} \begin{pmatrix} 0 & -1 & 0 \\ 0 & 0 & 1 \end{pmatrix} \begin{pmatrix} y_{1,r+1} \\ y_{2,r+1} \\ d_{r+1} \end{pmatrix} < \begin{pmatrix} 0 \\ 0 \end{pmatrix} \\ \begin{pmatrix} 0 & 0 & 1 \\ -d_3 & 0 & y_{1,3} \\ -\bar{D}^1_{3,4} & \bar{D}^2_{3,4} & \bar{D}^3_{3,4} \\ -\bar{D}^1_{5,4} & \bar{D}^2_{5,4} & \bar{D}^3_{5,4} \\ -\bar{D}^1_{5,6} & \bar{D}^2_{5,6} & \bar{D}^3_{5,6} \\ 0 & -d_6 & y_{2,6} \\ \vdots & \vdots & \vdots \\ 0 & -d_{r-1} & y_{2,r-1} \\ 0 & d_3 & 1 - y_{2,3} \\ \vdots & \vdots & \vdots \\ 0 & d_r & 1 - y_{2,r} \\ 0 & d_r & -y_{2,r} \end{pmatrix} \begin{pmatrix} y_{1,r+1} \\ y_{2,r+1} \\ d_{r+1} \end{pmatrix} \leq \begin{pmatrix} d_3 \\ -\bar{D}_{3,2,4} \\ -\bar{D}_{3,4,5} \\ -\bar{D}_{5,4,6} \\ -\bar{D}^1_{5,6} \\ -\bar{D}^1_{6,7} \\ \vdots \\ -\bar{D}^1_{r-1,r} \\ 0 \\ \vdots \\ 0 \\ d_r \end{pmatrix} \end{array}\right\} \qquad (5.2.5.2.9)$$

---

[167] The determinants are defined as in Section 5.2.5.1. Note that the system can be inconsistent (cf. Section 5.2.6).

Proof:[168] According to Theorem 4.4.2.1, problem (5.2.5.2.7) is a cycling example with simplex cycle C^{r+1} in (5.2.5.2.8) if and only if the following determinant inequality system is satisfied:

$$\left.\begin{array}{l}
D_{3,2},\quad D_{3,4}, D_{5,4}, D_{5,6}, D_{1,6}, D_{1,7}, \ldots, D_{1,r+1} > 0 \\
\bar{D}_{1,2,3} \begin{cases} < & 0 \\ \leq & \bar{D}_{1,2,\nu} \quad \forall \nu = 1, \ldots, r+1 \end{cases} \\
\bar{D}_{3,2,4} \begin{cases} < & 0 \\ \leq & \bar{D}_{3,2,\nu} \quad \forall \nu = 1, \ldots, r+1 \end{cases} \\
\bar{D}_{3,4,5} \begin{cases} < & 0 \\ \leq & \bar{D}_{3,4,\nu} \quad \forall \nu = 1, \ldots, r+1 \end{cases} \\
\bar{D}_{5,4,6} \begin{cases} < & 0 \\ \leq & \bar{D}_{5,4,\nu} \quad \forall \nu = 1, \ldots, r+1 \end{cases} \\
\bar{D}_{5,6,1} \begin{cases} < & 0 \\ \leq & \bar{D}_{5,6,\nu} \quad \forall \nu = 1, \ldots, r+1 \end{cases} \\
\bar{D}_{1,6,7} \begin{cases} < & 0 \\ \leq & \bar{D}_{1,6,\nu} \quad \forall \nu = 1, \ldots, r+1 \end{cases} \\
\bar{D}_{1,7,8} \begin{cases} < & 0 \\ \leq & \bar{D}_{1,7,\nu} \quad \forall \nu = 1, \ldots, r+1 \end{cases} \\
\qquad\vdots \\
\bar{D}_{1,r,r+1} \begin{cases} < & 0 \\ \leq & \bar{D}_{1,r,\nu} \quad \forall \nu = 1, \ldots, r+1 \end{cases} \\
\bar{D}_{1,r+1,2} \begin{cases} < & 0 \\ \leq & \bar{D}_{1,r+1,\nu} \quad \forall \nu = 1, \ldots, r+1 \end{cases}
\end{array}\right\} \qquad (5.2.5.2.10)$$

Since (5.2.5.2.5) is a cycling example, that system is satisfied which results from (5.2.5.2.10) by substituting r for $r+1$. Hence (5.2.5.2.10) is reducible to the subsystem

---

168 In the interest of brevity we confine ourselves to the essential steps.

$$\left.\begin{array}{r}
D_{1,r+1} > 0 \\
\bar{D}_{1,2,3} \leq \bar{D}_{1,2,r+1} \\
\bar{D}_{3,2,4} \leq \bar{D}_{3,2,r+1} \\
\bar{D}_{3,4,5} \leq \bar{D}_{3,4,r+1} \\
\bar{D}_{5,4,6} \leq \bar{D}_{5,4,r+1} \\
\bar{D}_{5,6,1} \leq \bar{D}_{5,6,r+1} \\
\bar{D}_{1,6,7} \leq \bar{D}_{1,6,r+1} \\
\vdots \\
\bar{D}_{1,r-1,r} \leq \bar{D}_{1,r-1,r+1} \\
\bar{D}_{1,r,r+1} \leq \bar{D}_{1,r,2} \\
\bar{D}_{1,r+1,2}\begin{cases} < & 0 \\ \leq & \bar{D}_{1,r+1,\nu} \quad \forall \nu = 3,\ldots,r \end{cases}
\end{array}\right\} \qquad (5.2.5.2.11)$$

Note that

$$\bar{D}_{1,r,2}\begin{cases} < 0 \\ \leq \quad \bar{D}_{1,r,\nu} \; \forall \nu = 1,\ldots,r \end{cases}$$

follows from the assumptions above.

Certain transformations of (5.2.5.2.11) (expansion of determinants by columns among others) yield (5.2.5.2.9).•

Using the above result and the notations

$$\left.S_0 = \begin{pmatrix}
0, & -1, & 0 \\
0, & 0, & 1 \\
0, & 0, & 1 \\
-d_3, & 0, & y_{1,3} \\
-\bar{D}^1_{3,4}, & \bar{D}^2_{3,4}, & \bar{D}^3_{3,4} \\
-\bar{D}^1_{5,4}, & \bar{D}^2_{5,4}, & \bar{D}^3_{5,4} \\
-\bar{D}^1_{5,6}, & \bar{D}^2_{5,6}, & \bar{D}^3_{5,6} \\
0, & d_3, & 1-y_{2,3} \\
0, & d_4, & 1-y_{2,4} \\
0, & d_5, & 1-y_{2,5} \\
0, & d_6, & 1-y_{2,6} \\
0, & d_6, & -y_{2,6}
\end{pmatrix}, \quad t_0 = \begin{pmatrix}
0 \\ 0 \\ d_3 \\ -\bar{D}_{3,2,4} \\ -\bar{D}_{3,4,5} \\ -\bar{D}_{5,4,6} \\ -\bar{D}^1_{5,6} \\ 0 \\ 0 \\ 0 \\ 0 \\ d_6
\end{pmatrix}\right\} \qquad (5.2.5.2.12)$$

(cf. (5.2.5.2.9)) we can list the algorithm in detail:

Input data:

Initial tableau

$$T = \begin{pmatrix} y_{1,1} & \cdots & y_{1,6} \\ y_{2,1} & \cdots & y_{2,6} \\ -d_1 & \cdots & -d_6 \end{pmatrix} \quad \text{in (5.2.5.3) (cf. (5.2.5.2.1))},$$

required column number N in (5.2.5.2.3),

column counter $r = 6$.

Set $S = S_0, t = t_0$ (cf. 5.2.5.2.12).

Step 1 (Addition of a column):

a) Determine a vector $(y_{1,r+1}, y_{2,r+1}, d_{r+1})^T$ with

$$S \begin{pmatrix} y_{1,r+1} \\ y_{2,r+1} \\ d_{r+1} \end{pmatrix} \begin{Bmatrix} < \\ \leq \end{Bmatrix} t$$

(cf. (5.2.5.2.9)).

b) Add a column to T, i.e. set

$$T = \left(T \;\middle|\; \begin{matrix} y_{1,r+1} \\ y_{2,r+1} \\ -d_{r+1} \end{matrix} \right).$$

c) Delete the last row of S and the last component of t.

d) Add 3 rows (components) to S (to t), i.e. set

$$S = \begin{pmatrix} & S & \\ \hline 0 & -d_r & y_{2,r} \\ 0 & d_{r+1} & 1 - y_{2,r+1} \\ 0 & d_{r+1} & -y_{2,r+1} \end{pmatrix}, \quad t = \begin{pmatrix} t \\ \hline -\bar{D}^1_{r,r+1} \\ 0 \\ d_{r+1} \end{pmatrix}$$

Set $r = r + 1$.

Step 2 (Test of stopping criterion):

If $r < N$, go to step 1.

Else: Print tableau

$$T = \begin{pmatrix} 1 & 0 & y_{1,3}, & \ldots, & y_{1,N} \\ 0 & 1 & y_{2,3}, & \ldots, & y_{2,N} \\ 0 & 0 & -d_3, & \ldots, & -d_N \end{pmatrix} \quad \text{– STOP}$$

Fig. 5.2.5.2.1 contains the corresponding flow-chart.

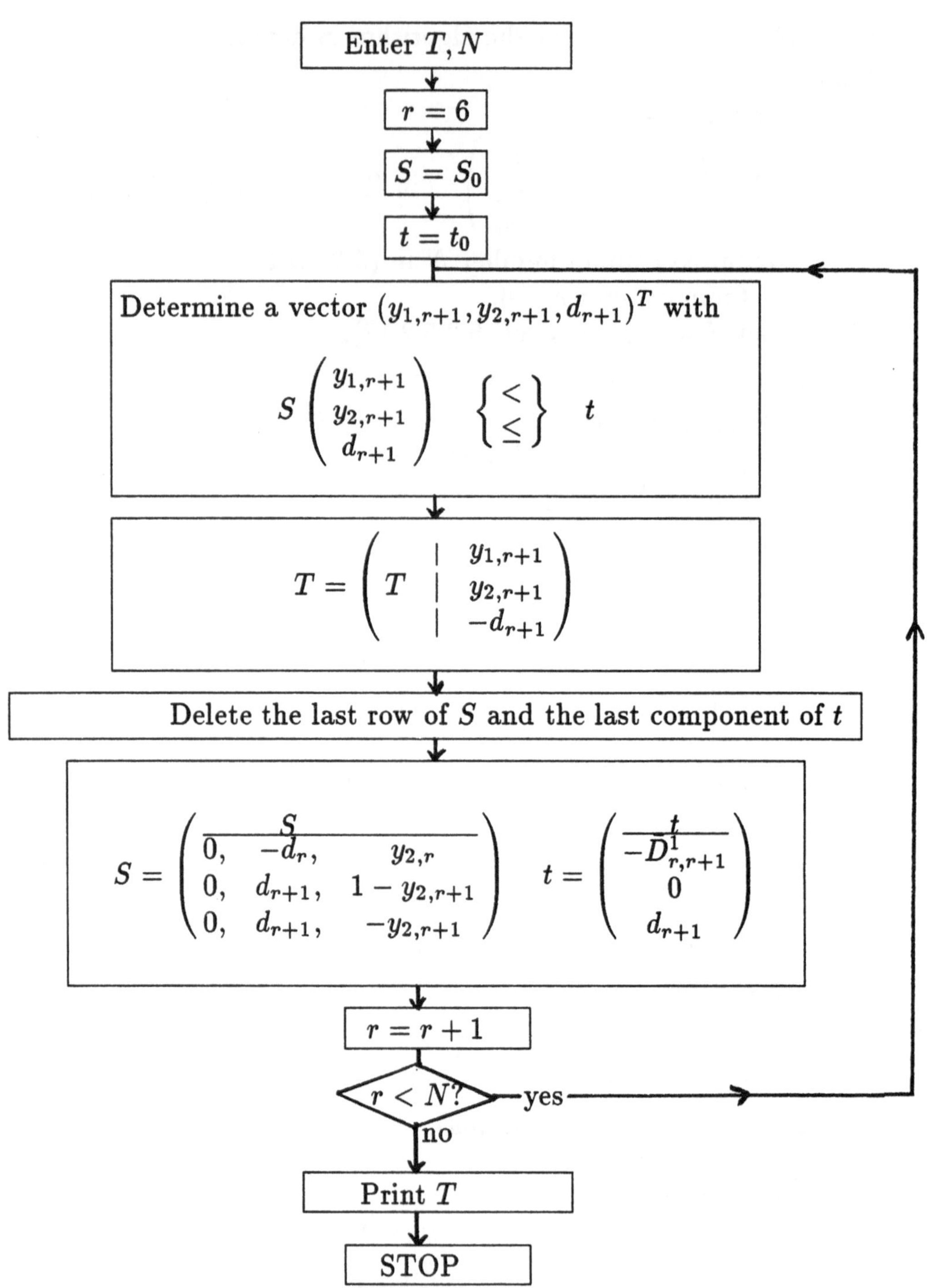

Fig. 5.2.5.2.1
Flow-chart for successive addition of columns

5.2.6 OPEN QUESTIONS IN CONNECTION WITH THE PRACTICAL PERFORMANCE OF THE PROCEDURES

Theoretical preparations of the above combined procedures end with a listing of their algorithms. However, certain questions remain open which can only be answered by extensive practical computer calculations.

By means of successive row modifications[169] (cf. Section 5.2.5.1) we can generate an arbitrary number of cycling examples, since the system of linear inequalities in Theorem 5.2.5.1.1 can never be inconsistent.[170] It remains to be tested in particular whether or for which initial data, tableaux occur repeatedly, or whether the generated tableau sequence converges to a limiting tableau under certain conditions.

In contrast to the above procedures, it is not insured that an unrestricted number of construction steps can be performed in the successive column addition (cf. Section 5.2.5.2). For certain cycling examples with two constraints, the procedure[171] was forced to terminate after some iterations, since system (5.2.5.2.9) was inconsistent, making the addition of another column impossible.

It should be proved by means of initial tableaux with different dimensions whether or after how many steps the procedure terminates. For this purpose the column addition must be programmed for cycling examples with arbitrary row numbers, where the initial tableau can be constructed as in Section 5.2.4.

If the forced termination after a small number of iterations should turn out to be the rule in column addition, the following modifications could possibly avoid the problem:

- After each column addition, any row or column *modifications* as in Sections 5.2.1 and 5.2.2 could be performed.

---

[169] In this connection certain programs were written by Götz (cf. Götz (1989)).

[170] This follows from the fact that the original row of T represents a solution of (5.2.5.1.3).

[171] Programs have been written by F. Geue and R. Götz.

- Based on a variant of Theorem 5.2.5.2.1 we could try to construct cycling examples with other simplex cycles, e.g. with a simplex cycle of the form

  $$\tilde{C}^r = (\{1,2\},\{3,2\},\{3,4\},\{5,4\},\{5,6\},\{7,6\},\{7,1\},\{7,2\},\\ \{8,2\},\ldots,\{r,2\},\{1,2\}).$$

  An initial cycling example for $\tilde{C}^7$ could be obtained from (5.2.5.1) analogously to Section 5.2.3 (with $C = (\{1,2\}, \{2,3\}, \{3,4\}, \{4,5\}, \{5,6\}, \{1,6\}, \{1,2\}))$.

5.3 ON THE CONSTRUCTION OF GENERAL CYCLING EXAMPLES

We can construct a general cycling example (i.e. one in which the right-hand side differs from zero) by the addition of appropriate rows and columns to the initial tableau of a reduced cycling example (cf. Sections 4.1 and 5.2.4). Let

$$\left.\begin{array}{llllllll}
\max z = & d_1x_1+ & \ldots+ & d_4x_4 & & & & \\
\text{s.t.} & & & & & & & \\
 & y_{1,1}x_1+ & \ldots+ & y_{1,4}x_4+ & x_5 & & & =0 \\
 & y_{2,1}x_1+ & \ldots+ & y_{2,4}x_4+ & & x_6 & & =0 \\
 & & & & & & x_1,\ldots,x_6 & \geq 0
\end{array}\right\} \quad (5.3.1)$$

be a reduced cycling example with simplex cycle

$$C = (I_1,\ldots,I_k,I_1) \qquad (5.3.2)$$

(cf. (5.2.5.1), (5.2.5.2)).

From this we obtain the following general cycling example with one additional constraint and one additional variable:

$$\left.\begin{array}{lllllllll}
\max z = & d_1x_1+ & \ldots+ & d_4x_4 & & & & & \\
\text{s.t.} & & & & & & & & \\
 & y_{1,1}x_1+ & \ldots+ & y_{1,4}x_4+ & x_5 & & & & =0 \\
 & y_{2,1}x_1+ & \ldots+ & y_{2,4}x_4+ & & x_6 & & & =0 \\
 & y_{3,1}x_1+ & \ldots+ & y_{3,4}x_4+ & & & x_7 & & =y_3 \\
 & & & & & & & x_1,\ldots,x_7 & \geq 0
\end{array}\right\} \quad (5.3.3)$$

The simplex cycle is

$$C' = (I_1 \cup \{7\}, \dots, I_k \cup \{7\}, I_1 \cup \{7\}). \tag{5.3.4}$$

In (5.3.3) the coefficients $y_{3,1}, \dots, y_{3,4}$ and the component y_3 can be chosen *arbitrarily* ($y_3 > 0$; cf. the remarks following (5.3.7)). For the purposes of illustration let (5.3.1), (5.3.2) be of the special form (5.2.1.1), (5.2.1.2), respectively.

Then the general cycling example (5.3.3) can e.g. have the form

$$\left.\begin{array}{lllllllll}
\max z = & \frac{3}{4}x_1 - & 150x_2 + & \frac{1}{50}x_3 - & 6x_4 & & & & \\
\text{s.t.} & & & & & & & & \\
& \frac{1}{4}x_1 - & 60x_2 - & \frac{1}{25}x_3 + & 9x_4 + & x_5 & & & = 0 \\
& \frac{1}{2}x_1 - & 90x_2 - & \frac{1}{50}x_3 + & 3x_4 + & & x_6 & & = 0 \\
& -x_1 + & 5x_2 - & 6x_3 + & x_4 + & & & x_7 & = 2 \\
& & & & & & & x_1, \dots, x_7 & \geq 0
\end{array}\right\} \tag{5.3.5}$$

where the simplex cycle is

$$C' = (\{1,2,7\}, \{2,3,7\}, \{3,4,7\}, \{4,5,7\}, \{5,6,7\}, \{1,6,7\}, \{1,2,7\}) \tag{5.3.6}$$

The associated tableaux are listed in Tab. 5.3.1. The above cycle C' represents a cycle in the degenerate vertex $x^0 = (0,0,0,0)^T$ of the solution set X of

$$\left.\begin{array}{llllll}
\max z = & \frac{3}{4}x_1 - & 150x_2 + & \frac{1}{50}x_3 - & 6x_4 & \\
\text{s.t.} & & & & & \\
& \frac{1}{4}x_1 - & 60x_2 - & \frac{1}{25}x_3 + & 9x_4 & \leq 0 \\
& \frac{1}{2}x_1 - & 90x_2 - & \frac{1}{50}x_3 + & 3x_4 & \leq 0 \\
& -x_1 + & 5x_2 - & 6x_3 + & x_4 & \leq 2 \\
& & & & x_1, \dots, x_4 & \geq 0
\end{array}\right\} \quad (5.3.7)^{172}$$

Tab. 5.3.1 illustrates that $y_{3,1}, \dots, y_{3,4}$ and y_3 in (5.3.3) can be chosen arbitrarily with $y_3 > 0$. Because of $y_3 > 0$, the pivot element

[172] The corresponding canonical form is (5.3.5).

Tab. 5.3.1
Tableaux of the cycling example (5.3.5) or 5.3.7) with simplex cycle C' in (5.3.6)

	1	2	3	4	5	6	7	x_B
5	$\frac{1}{4}$	- 60	$-\frac{1}{25}$	9	1	0	0	0
6	$\frac{1}{2}$	-90	$-\frac{1}{50}$	3	0	1	0	0
7	-1	5	-6	1	0	0	1	$\overline{2}$
Δz_j	$-\frac{3}{4}$	150	$-\frac{1}{50}$	6	0	0	0	0
1	1	-240	$-\frac{4}{25}$	36	4	0	0	0
6	0	30	$\frac{3}{50}$	-15	-2	1	0	0
7	0	-235	$-\frac{154}{25}$	37	4	0	1	$\overline{2}$
Δz_j	0	-30	$-\frac{7}{50}$	33	3	0	0	0
1	1	0	$\frac{8}{25}$	−84	-12	8	0	0
2	0	1	$\frac{1}{500}$	$-\frac{1}{2}$	$-\frac{1}{15}$	$\frac{1}{30}$	0	0
7	0	0	$-\frac{569}{100}$	$-\frac{161}{2}$	$-\frac{35}{3}$	$\frac{47}{6}$	1	$\overline{2}$
Δz_j	0	0	$-\frac{2}{25}$	18	1	1	0	0
3	$\frac{25}{8}$	0	1	$-\frac{525}{2}$	$-\frac{75}{2}$	25	0	0
2	$-\frac{1}{160}$	1	0	$\frac{1}{40}$	$\frac{1}{120}$	$-\frac{1}{60}$	0	0
7	$\frac{569}{32}$	0	0	$-\frac{12593}{8}$	$-\frac{5401}{24}$	$\frac{1801}{12}$	1	$\overline{2}$
Δz_j	$\frac{1}{4}$	0	0	-3	-2	3	0	0
3	$-\frac{125}{2}$	10500	1	0	50	-150	0	0
4	$-\frac{1}{4}$	40	0	1	$\frac{1}{3}$	$-\frac{2}{3}$	0	0
7	$-\frac{1503}{4}$	62965	0	0	$\frac{899}{3}$	$-\frac{2698}{3}$	1	$\overline{2}$
Δz_j	$-\frac{1}{2}$	120	0	0	-1	1	0	0
5	$-\frac{5}{4}$	210	$\frac{1}{50}$	0	1	-3	0	0
4	$\frac{1}{6}$	-30	$-\frac{1}{150}$	1	0	$\frac{1}{3}$	0	0
7	$-\frac{7}{6}$	35	$-\frac{899}{150}$	0	0	$-\frac{1}{3}$	1	$\overline{2}$
Δz_j	$-\frac{7}{4}$	330	$\frac{1}{50}$	0	0	-2	0	0

The next pivot step generates the initial tableau.

Legend: Tableau associated with basis B

	column indices
indices of the basis B	$B^{-1}\bar{Y}$
Δz_j	$c_B^T B^{-1}\bar{Y} - \bar{d}$

The subtableaux in bold type correspond to the tableaux of (5.3.1), (5.3.2) (cf. also Tab. 4.1.2).

is always chosen from the subtableau in bold type, although the third row is changed in each step.

An example of simplex cycling in a degenerate vertex $\tilde{x}^0 \neq 0$ can be easily obtained from (5.3.5) as follows: Pivoting in the initial tableau of Tab. 5.3.1 with pivot element $y_{3,2} = 5$ yields the tableau (after exchanging the columns 2 and 7 and "re-indexing"):

	1	2	3	4	5	6	7	
5	$-\frac{47}{4}$	12	$-\frac{1801}{25}$	21	1	0	0	24
6	$-\frac{35}{2}$	18	$-\frac{5401}{50}$	21	0	1	0	36
7	$-\frac{1}{5}$	$\frac{1}{5}$	$-\frac{6}{5}$	$\frac{1}{5}$	0	0	1	$\frac{2}{5}$
Δz_j	$\frac{117}{4}$	-30	$\frac{8999}{50}$	-24	0	0	0	-60

(5.3.8)

Tableau (5.3.8) is the initial tableau of the following linear optimization problem:[173]

$$\left.\begin{array}{llllll} \max z = & -\frac{117}{4}x_1 + & 30x_2 - & \frac{8999}{50}x_3 + & 24x_4 & \\ \text{s.t.} & & & & & \\ & -\frac{47}{4}x_1 + & 12x_2 - & \frac{1801}{25}x_3 + & 21x_4 & \leq 24 \\ & -\frac{35}{2}x_1 + & 18x_2 - & \frac{5401}{50}x_3 + & 21x_4 & \leq 36 \\ & -\frac{1}{5}x_1 + & \frac{1}{5}x_2 - & \frac{6}{5}x_3 + & \frac{1}{5}x_4 & \leq \frac{2}{5} \\ & & & & x_1, \ldots, x_4 & \geq 0 \end{array}\right\} \tag{5.3.9}$$

Now problem (5.3.9) represents an example of simplex cycling in the degenerate vertex $\tilde{x}^0 = (0,2,0,0)^T$ of the solution set $\tilde{X}$ of (5.3.9). Pivoting in the initial tableau (5.3.8) of (5.3.9) with pivot element $y_{3,2} = \frac{1}{5}$ yields a tableau of the degenerate vertex $\tilde{x}^0 = (0,2,0,0)^T$. The corresponding index set is $\{2,5,6\}$. Beginning with this tableau, cycling in $\tilde{x}^0$ can occur,where the simplex cycle is

$$C'' = (\{2,5,6\},\{1,2,6\},\{1,2,7\},\{2,3,7\},\{2,3,4\},\{2,4,5\},\{2,5,6\}). \tag{5.3.10}$$

---

173 We assume that the objective function value in (5.3.8) equals zero.

Except for the objective function values and the exchange of columns 2 and 7, the tableaux of this cycling example coincide with those in Tab. 5.3.1.

Summary of Chapter 5

A sufficient stock of cycling examples is necessary for clarifying practical questions, in particular for testing the efficiency of anticycling rules (cf. Section 5.1). Since the construction of cycling examples is not yet to be found in literature, we present a certain type of construction procedures for cycling examples in Chapter 5.

The "determinant approach" (cf. Ch. 4) implies that a (reduced) linear optimization problem (RLOP) is a cycling example if and only if its coefficients satisfy a certain *nonlinear* determinant inequality system. Thus in principle, cycling examples could be generated using any method for solving nonlinear inequality systems (cf. Appendix C). However, on grounds of the complexity of determinant inequality systems, such a method is not suitable.

Therefore we have developed specific procedures for constructing (reduced) cycling examples. The basic idea is to transform a given cycling example by a sequence of elementary construction steps, in each of which a single row (column) is modified or one row (column) is added to the tableau. The advantage of such a step-by-step method is that "only" a *linear* inequality system must be solved in each step. Diverse construction procedures result from the possibilities of combining construction steps.

For practical applications, on the one hand it is important to make as many cycling examples available as possible. On the other hand, cycling examples with dimensions as large as possible are needed. These considerations led to the following two combined procedures:

1) *Successive modification of rows*
   Beginning with the initial tableau of a given cycling example, all rows are modified repeatedly. In this way any number of cycling examples can be generated, but all of them have the same dimension.

2) *Successive addition of columns*
   Step by step we add another column to the initial tableau. In principle, cycling examples with any desired column dimension could be generated by this mean (cf. below).

In order to evaluate the usefulness of the above procedures definitively, extensive computer tests are necessary which could not be performed within the scope of the present publication. Initial tests of procedure 2) by means of cycling examples with two constraints showed that it is often forced to terminate after few iterations, since the respecting inequality system can be inconsistent (cf. Section 5.2.6). However, it remains to be seen whether this problem occurs using initial tableaux with higher row numbers, in which case one could try to overcome the problem by certain variations in the column addition. For example a row or column modification could be inserted between any successive column additions. Apart from the above procedures 1) and 2), other combinations of construction steps in Sections 5.2.1 – 5.2.4 are conceivable; e.g. rows and columns could be added alternately, in order to increase row dimension as well as column dimension.

Independently of the results of the above computer tests, Chapter 5 provides a theoretical basis for constructing cycling examples of arbitrary dimensions.

APPENDIX

A. FOUNDATIONS OF LINEAR ALGEBRA AND THE THEORY OF CONVEX POLYTOPES

This section contains the notations and statements from linear algebra[174] and the theory of convex polytopes[175] used in the present publication.

Definition A.1: A set $M \subset \mathbb{R}^n$ is called *convex* if and only if for each two points $x, y \in M$ the closed segment $V(x,y) := \{z \in \mathbb{R}^n | z = x + \lambda(y-x), \lambda \in [0,1]\}$ is contained in M, i.e. if

$$x, y \in M \Rightarrow V(x,y) \subset M.$$

holds.

Remark A.2: The intersection of any convex sets is also convex.

Definition A.3: A subset H of $\mathbb{R}^n$ with

$$H = H_{a,b} = \{x \in \mathbb{R}^n | a^T x \leq b\} \quad (a \in \mathbb{R}^n, b \in \mathbb{R})$$

is called a *(closed) halfspace* of $\mathbb{R}^n$. A set of the form

$$E = E_{\bar{a},\bar{b}} = \{x \in \mathbb{R}^n | \bar{a}^T x = \bar{b}\} \quad (\bar{a} \in \mathbb{R}^n, \bar{b} \in \mathbb{R})$$

is called a *hyperplane* of $\mathbb{R}^n$. The hyperplane $E_{a,b}$ is called the *constraint-hyperplane belonging to* $H_{a,b}$.

Remark A.4: Obviously any halfspace is a convex set.

Definition A.5: The intersection of a finite family of closed halfspaces of $\mathbb{R}^n$ is called a *convex polyhedral set* (cf. the remarks above). A bounded convex polyhedral set is called a *convex polytope.*

---

174 Cf. e.g. Kowalsky (1975).

175 Cf. Grünbaum (1967) and Kruse (1986, Appendix A).

Remark A.6: Obviously the solution set of any system of linear inequalities is a convex polyhedral set.

Definition A.7 A point set $K \subset \mathbb{R}^n$ is called a *cone (with apex 0)* if the following holds:

$$x \in K \Rightarrow \lambda x \in K \quad \forall x \in \mathbb{R}^n$$

$$\forall \lambda \in \mathbb{R}, \lambda \geq 0$$

Lemma A.8: The solution set of a homogeneous system of linear inequalities

$$Ax \leq 0$$

$(A \in \mathbb{R}^{m \times n}, x \in \mathbb{R}^n)$ is a convex cone.

Proof: From $Ax \leq 0$ follows $A(\lambda x) = \lambda(Ax) \leq 0$ for $\lambda \geq 0$. The convexity follows from the above remarks. •

Definition A.9: Let $X \subset \mathbb{R}^n$ be a convex polyhedral set. A point $x \in X$ is called a *vertex* of X if x does *not* belong to the interior of any closed segment, i.e. if

$$x \in V(y, z) \Rightarrow (x = y \text{ or } x = z) \quad \forall y, z \in X$$

holds (cf. Def. A.1).

Definition A.10: A point x of a convex polyhedral set $X \subset \mathbb{R}^n$ is called an *interior point* of X if a (sufficiently small) neighbourhood of x is contained in X, i.e. if

$$|x' - x| \leq \varepsilon \Rightarrow x' \in X$$

holds for a sufficiently small $\varepsilon \in \mathbb{R}, \varepsilon > 0$. The set of all interior points of X is called the *interior of* X and is denoted by int X.

Definition A.11: A subset $U \subset \mathbb{R}^n$ is called a *subspace* of $\mathbb{R}^n$ if

$$x, y \in U \Rightarrow x + y \in U$$

$$\alpha \in \mathbb{R}, x \in U \Rightarrow \alpha x \in U$$

holds. A set of the form

$$x + U := \{x + u | u \in U\},$$

where U is a subspace of $\mathbb{R}^n$, is called an *affine subspace of* $\mathbb{R}^n$.

Remark A.12: Obviously any subspace of $\mathbb{R}^n$ is an affine subspace. In particular, every hyperplane is an affine subspace.

Example A.13: The subspaces of $\mathbb{R}^3$ are the point $0 \in \mathbb{R}^3$, the straight lines and planes, containing 0, and $\mathbb{R}^3$ itself. The affine subspaces of $\mathbb{R}^3$ are *all* points of $\mathbb{R}^3$, *all* straight lines and planes in $\mathbb{R}^3$, and $\mathbb{R}^3$ itself.

Definition A.14: For any vectors $x^0, x^1, \ldots, x^k \in \mathbb{R}^n$ we define $U(x^1, \ldots, x^k)$ and $U(x^0; x^1, \ldots, x^k)$ by:

$$\begin{aligned} U(x^1, \ldots, x^k) &:= \text{span } \{x^1, \ldots, x^k\} \\ &:= \{\alpha_1 x^1 + \ldots, +\alpha_k x^k | \alpha_1, \ldots, \alpha_k \in \mathbb{R}\} \end{aligned}$$

and

$$\begin{aligned} U(x^0; x^1, \ldots, x^k) &:= x^0 + U(x^1, \ldots, x^k) \\ &= \{x^0 + u | u \in U(x^1, \ldots, x^k)\}. \end{aligned}$$

Remark A.15: The point sets $U(x^1, \ldots, x^k)$ and $U(x^0; x^1, \ldots, x^k)$ represent subspaces or affine subspaces, respectively.

Definition A.16: The subspace $U(x^1, \ldots, x^k)$is called the *subspace generated by the vectors* $x^1, \ldots, x^k$; the affine subspace $U(x^0; x^1, \ldots, x^k)$ is called a *subspace, parallel with* $U(x^1, \ldots, x^k)$.

B. FOUNDATIONS OF GRAPH THEORY

In the following we present the necessary concepts and interrelations of graph theory which are drawn from standard literature in the main.[176]

Definition B.1: A *graph* is an ordered pair $G = (V, E)$, where V is a finite, nonempty set and E is a set of two-element subsets of V. The sets V, E are called *node set* and *edge set*, respectively; $v \in V$ is called a *node* of G and $\{v_1, v_2\} \in E$ is called an *edge* of G. Two nodes $v_1, v_2 \in V$ are said to be *adjacent (in G)* or *joined by an edge (in G)* if $\{v_1, v_2\} \in E$. Two edges are said to be *adjacent* if they have a common node.

A node $v \in V$ and an edge $\{v_1, v_2\} \in E$ are said to be *incident (in G)* if $v = v_1$ or $v = v_2$. The number of edges, incident to a node v is called the *degree of* v, denoted by $g(v)$.

Definition B.2: Two graphs $G = (V, E)$ and $G^* = (V^*, E^*)$ are said to be *isomorphic* (written $G \cong G^*$) if there is a mapping $\varphi : V \to V^*$ such that

$$\{v_1, v_2\} \in E \Leftrightarrow \{\varphi(v_1), \varphi(v_2)\} \in E^*.$$

Definition B.3: Let $G = (V, E)$ be a graph. A *subgraph* of G is a graph $G^* = (V^*, E^*)$ such that $V^* \in V$ and $E^* \in E$. If W is a set of nodes in G, then the *subgraph induced by* W (written $< W >$) is the subgraph of G obtained by taking the vertices in W and joining those pairs of vertices in W which are joined in G. Two subgraphs $G_1 = (V, E_1)$, $G_2 = (V, E_2)$ of G are called *complementary graphs (in G)* if $E_1 \cap E_2 = \emptyset$ and $E_1 \cup E_2 = E$.

Definition B.4: Let $v_1, \ldots, v_n$ be a finite sequence of nodes in G. Moreover let

---

[176] Cf. e.g. Behzad/Chartrand (1971), Beineke/Wilson (1978), Harary (1974), Wagner (1970) and Kruse (1986, Appendix B).

$$K_i := \{v_i, v_{i+1}\} \in E \quad \text{for} \quad i = 1, \ldots, n-1,$$

be pairwise distinct edges of G. Then the subgraph of G, consisting of the nodes $v_1, \ldots, v_n$ and the edges $K_1, \ldots, K_{n-1}$ is called a *trail from v_1 to v_n (in G)*. It is denoted by $(v_1, \ldots, v_n)$.

The number $n-1$ is called the *length* of $(v_1, \ldots, v_n)$. The trail $(v_1, \ldots, v_n)$ is called a *path* if the nodes $v_1, \ldots, v_n$ are pairwise distinct.

A trail $(v_1, \ldots, v_n, v_1)$ is called a *cycle* if $(v_1, \ldots, v_n)$ is a path $(n \geq 2)$.

Two paths $W = (v_1, \ldots, v_n)$ and $W' = (v_1, v_2', \ldots, v_{n-1}', v_n)$ from v_1 to v_n $(n \geq 3)$ are called *(node-) disjoint* if they have no nodes in common except v_1 and v_n; W and W' are called *edge-disjoint* if they have no edge in common.

Remark B.5: Obviously any two node-disjoint paths are edge-disjoint. The reverse assertion does not hold.

Definition B.6: Let v, w be two nodes of a graph $G = (V, E)$. The length of a shortest path from v to w, denoted by $d(v, w)$, is called the distance between v and w $(d(v, v) := 0)$. The diameter of G, denoted by $d(G)$, is defined as the largest distance between two vertices in G, i.e.

$$d(G) = \max \{d(v, w) | v, w \in V\}.$$

Definition B.7:[177] A graph G is said to be *n-node connected (n-edge connected)* if each two distinct nodes of G are connected by n pairwise node-disjoint (edge-disjoint) paths. In abbreviated form we say *n-connected* instead of n-node connected and *connected* instead of 1-connected. The *node-connectivity* of G (*connectivity* for short), denoted by ω, is defined to be the largest value of n for which G is n-connected. Analogously the *edge-connectivity*, denoted by ω', is defined to be the largest number of n' for which G is n'-edge connected.

177 To simplify matters, the definition here differs from the "usual" definition of connectivities. However, the latter is equivalent to the above definition (cf. Behzad/Chartrand (1971, Theorems 10.5 and 10.7)).

Example B.8: The graph G in Fig. B.1 has node-connectivity $\omega = 1$ and edge-connectivity $\omega^{'} = 2$. Obviously each two nodes of G are connected by two edge-disjoint paths. However, each two paths between the nodes 1 and 7 have node 4 in common, i.e. they are *not* node-disjoint.

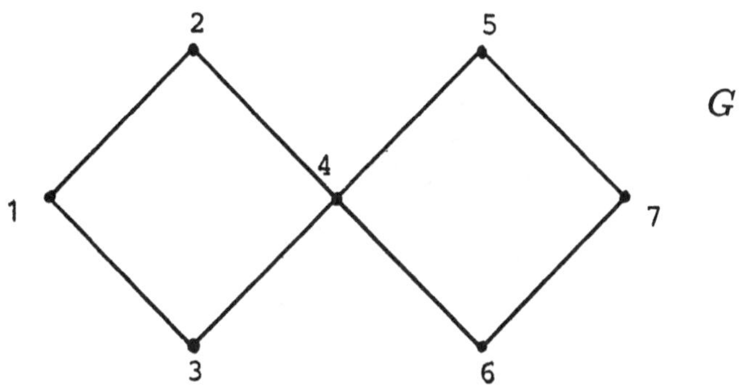

Fig. B.1
Graph with $\omega = 1$ and $\omega^{'} = 2$

Interrelations between connectivities and minimum degree are as follows (cf. Behzad/Chartrang (1971, Theorem 10.1) and Remark B.5):

Lemma B.9: Let G be a graph and let δ denote the minimum degree of G. Then

$$\omega \leq \omega^{'} \leq \delta.$$

Definition B.10: Let $G = (V, E)$ be a graph. It is said to be *complete* if each two nodes of G are adjacent. The graph G is called *r-partite* if a partition

$$V = V_1 \cup \ldots \cup V_r \qquad (B.1)$$

of the node set V exists, such that every edge of G joins nodes of *different* components of (B.1). If each two nodes of different components of (B.1) are joined by an edge, G is said to be a *complete r-partite graph.* For $|V_i| = p_i$ $(i = 1, \ldots, r)$ the complete r-partite graph is uniquely determined; it is denoted by $K(p_1, \ldots, p_r)$.

Example B.11: The complete 3-partite graph $K(2,2,3)$ is illustrated in Fig. B.2.

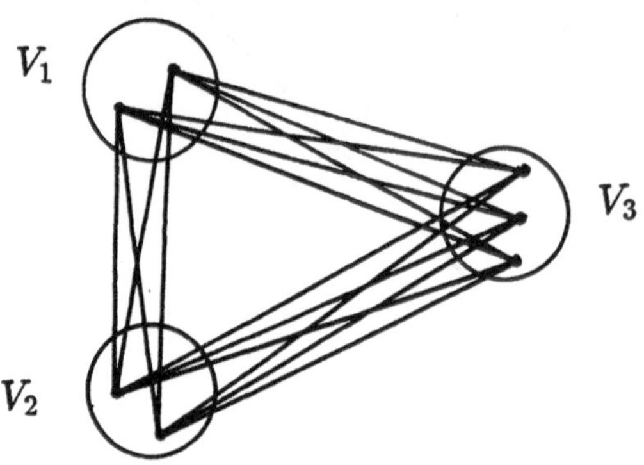

Fig. B.2
Representation of $K(2,2,3)$

Definition B.12: Let G be a graph. The *line graph* of G, denoted by $L(G)$, is defined as follows:

The nodes of $L(G)$ correspond to the edges of G, where two nodes of $L(G)$ are joined if and only if the corresponding edges of G are adjacent (cf. Def. B.1).

Example B.13: Fig. B.3 illustrates the definition above.

Definition B.14: Given a graph G and two different edges K, K' of G. A sequence of pairwise disjoint edges

$$(K = K_1, K_2, \ldots, K_n = K^{'}),$$

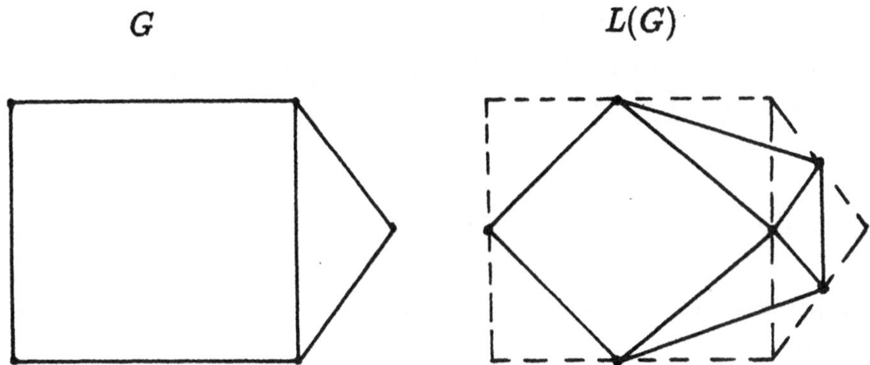

Fig. B.3
Construction of the line graph $L(G)$ for a given graph G

where K_i and K_{i+1} are adjacent for $(i = 1, \ldots, n-1)$, is called an *edge path* from K to K'.

Remark B.15: Edge paths in $L(G)$ correspond biuniquely to paths in G.

C. PROBLEMS IN THE SOLUTION OF DETERMINANT INEQUALITY SYSTEMS[178]

Let us now illustrate problems in solving (nonlinear) determinant inequality systems which have led to the development of successive construction procedures for cycling examples (cf. the introductory representations in Section 5.2).

Even in the case of small dimensions of $\bar{Y}$ and $\bar{d}$ in (5.2.1), the determinant inequality system (a) - (d) in Theorem 4.4.2.1 contains numerous variables and inequalities. In the "simplest" case with

$$\bar{Y} = \begin{pmatrix} y_{1,1}, & \ldots, & y_{1,4}, & 1, & 0 \\ y_{2,1}, & \ldots, & y_{2,4}, & 0, & 1 \end{pmatrix}$$

and

$$\bar{d} = (d_1, \ldots, d_4, 0, 0)$$

(a) - (d) has the form (5.2.1.4) (cf. Section 5.2.1). The latter is transformable into

$$\left.\begin{array}{ll} p_i(y_{1,1}, \ldots, y_{1,4}, y_{2,1}, \ldots, y_{2,4}, d_1, \ldots, d_4) & \geq 0 \\ \quad (i = 1, \ldots, 18) & \\ p_i(y_{1,1}, \ldots, y_{1,4}, y_{2,1}, \ldots, y_{2,4}, d_1, \ldots, d_4) & > 0 \\ \quad (i = 19, \ldots, 29) & \end{array}\right\} \qquad (C.1)$$

where $p_1, \ldots, p_{29}$ are polynomials in the 12 variables $y_{1,1}, \ldots, y_{1,4}$, $y_{2,1}, \ldots, y_{2,4}, d_1, \ldots, d_4$. Ten of these polynomials have degree 3, twelve have degree 2 and seven have degree 1.

The computational effort in the solution of a nonlinear system (C.1) depends essentially on the number of so-called "cells", increasing with

---

[178] The representations of this section are partially based on discussions with Prof. Möller, Fernuniversität Hagen, to whom I would like to express my gratitude here.

the number of inequalities.[179] For illustration of the concept "cell" we begin with considering a single inequality of the form

$$p(x,y) \leq 0 \tag{C.2}$$

where p is assumed to be a polynomial of the second degree with variables x and y. The point set $K = \{\binom{x}{y} \in I\!R^2 | p(x,y) = 0\}$ consists of one or two curves K_i in $I\!R^2$ (cf. Example C.1). The open sets bordered by these curves are said to be the *cells*[180] of (C.2).

As the following example shows, an inequality of the form (C.2) partitions the space $I\!R^2$ into two, three or four cells.[181]

Example C.1: Let the polynomial p in (C.2) be of the special form

a) $p(x,y) = y + \frac{1}{2}x^2 - 5$
b) $p(x,y) = -y^2 + x^2 + 3x$
c) $p(x,y) = xy$

The respective cells are illustrated in Fig. C.1.

Let us now consider a system with two inequalities

$$\left.\begin{array}{ll} p_1(x,y) & \leq 0 \\ p_2(x,y) & \leq 0 \end{array}\right\} \tag{C.3}$$

where p_1, p_2 are again polynomials of the second degree with variables x and y. Then the point set

$$K = \{\begin{pmatrix} x \\ y \end{pmatrix} \in I\!R^2 | p_i(x,y) = 0 \text{ for at least one } i \in \{1;2\}\}$$

---

[179] For a fixed number of variables the computational effort in constructing the CAD of (C.1) (cf. the representations following (C.5)) depends polynomially on the parameters of (C.1) (cf. Arnon et al. (1984:866)).

[180] We use the concept "cell" in a simplified sense. The exact definition (cf. Arnon et al. (1984, Section 2)) would be beyond the scope of this publication. Our intention here is only to illustrate the dependence between the numbers of inequalities and cells of a system, for which the above definition is suitable.

[181] If p in (C.2) is a polynomial in n variables, the space $I\!R^n$ is partitioned into cells.

a) $p(x,y) = y + \frac{1}{2}x^2 - 5$

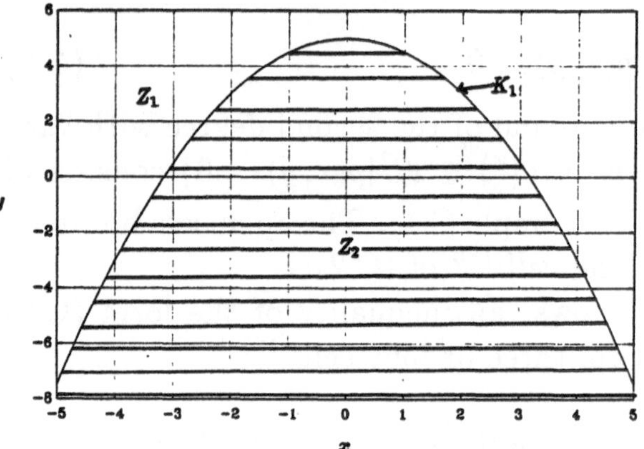

$K_1 = \{\binom{x}{y} | y = 5 - \frac{1}{2}x^2\}$

$Z_1 = \{\binom{x}{y} | y > 5 - \frac{1}{2}x^2\}$

$Z_2 = \{\binom{x}{y} | y < 5 - \frac{1}{2}x^2\}$

b) $p(x,y) = -y^2 + x^2 + 3x$

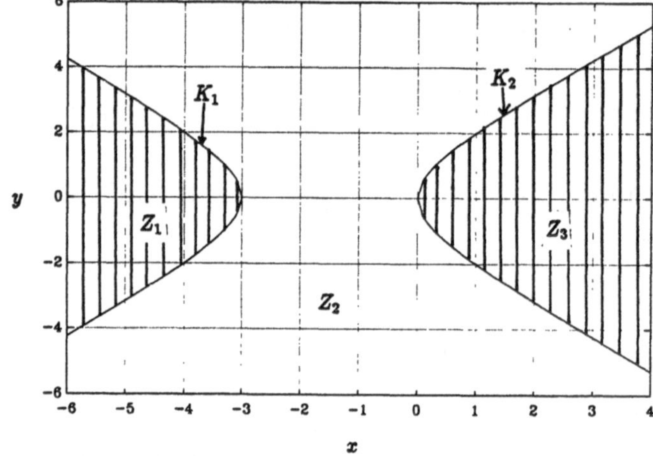

$K_1 = \{\binom{x}{y} | y^2 = x^2 + 3x, x \leq -3\}$

$K_2 = \{\binom{x}{y} | y^2 = x^2 + 3x, x \geq 0\}$

$Z_1 = \{\binom{x}{y} | y^2 < x^2 + 3x, x < -3\}$

$Z_2 = \{\binom{x}{y} | y^2 > x^2 + 3x\}$

$Z_3 = \{\binom{x}{y} | y^2 < x^2 + 3x, x > 0\}$

Fig. C.1
Representation of the cells Z_i of (C.2)
for Example C.1

c) $p(x, y) = xy$

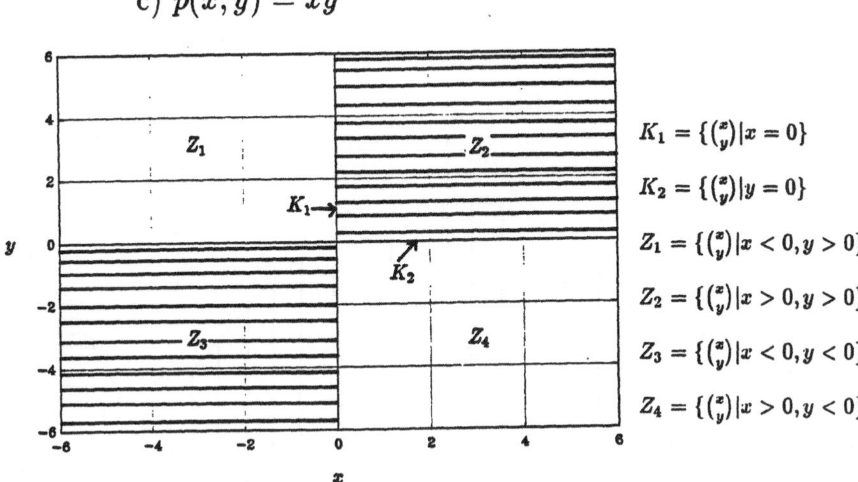

Legend: Z_i – i-th cell $\qquad$ K_i – i-th curve of the point set K

Fig. C.1
Representation of the cells Z_i of (C.2) for Example C.1 (continued)

consists of maximal four curves K_i, partitioning the space $\mathbb{R}^2$ into cells.

Example C.2: Let the polynomials p_1, p_2 in (C.3) be of the following form:

a) $p_1(x, y) = y - x^2 + 4$
$\quad p_2(x, y) = y + x^2 - 4$

b) $p_1(x, y) = -y^2 + x^2 + 3x$
$\quad p_2(x, y) = y + 0,3x^2 - 5$

c) $p_1(x, y) = x^2 - 16$
$\quad p_2(x, y) = y^2 - 16$

The respective cells are illustrated in Fig. C.2 - C.4.

As the above examples show, the cell number increases considerably with every further inequality. If we e.g. change from the inequality

$$p_1(x, y) = -y^2 + x^2 + 3x \tag{C.4}$$

to the system

$$\left.\begin{aligned} p_1(x, y) &= -y^2 + x^2 + 3x \\ p_2(x, y) &= y + 0,3x^2 - 5 \end{aligned}\right\} \tag{C.5}$$

$p_1(x,y) = y - x^2 + 4$

$p_2(x,y) = y + x^2 - 4$

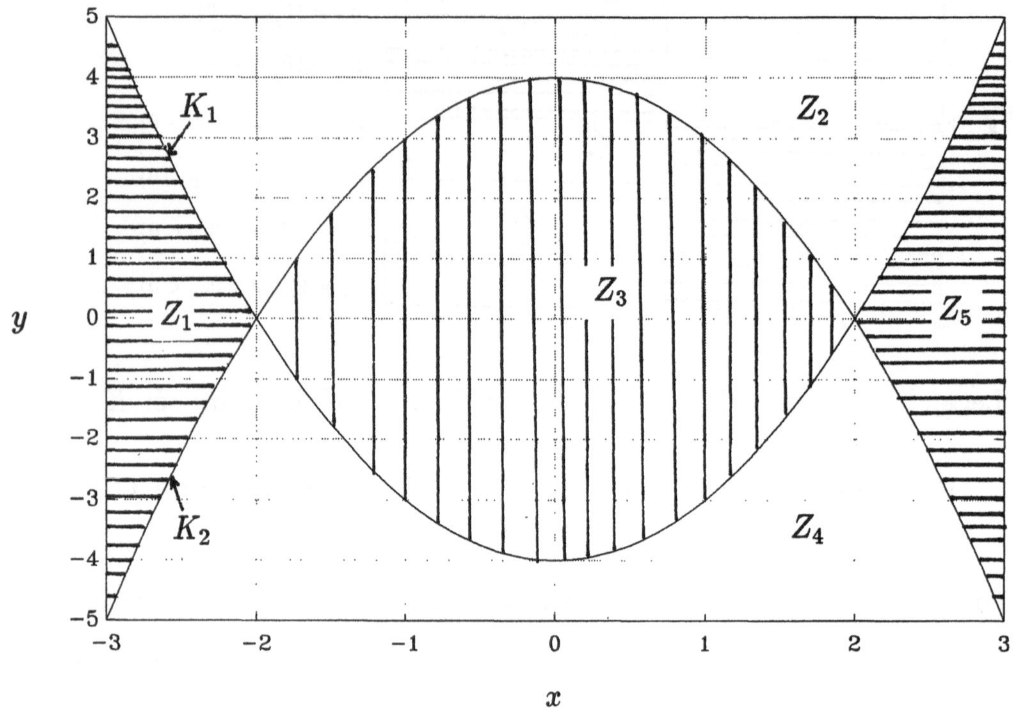

$K_1 = \{\binom{x}{y} | y = x^2 - 4\}$
$K_2 = \{\binom{x}{y} | y = -x^2 + 4\}$
$Z_1 = \{\binom{x}{y} | -x^2 + 4 < y < x^2 - 4, x < -2\}$
$Z_2 = \{\binom{x}{y} | -x^2 + 4 < y, x^2 - 4 < y\}$
$Z_3 = \{\binom{x}{y} | x^2 - 4 < y < -x^2 + 4\}$
$Z_4 = \{\binom{x}{y} | -x^2 + 4 > y, x^2 - 4 > y\}$
$Z_5 = \{\binom{x}{y} | -x^2 + 4 < y < x^2 - 4, x > 2\}$

Legend: Z_i – i-th cell

K_i – i-th curve of the point set K

Fig. C.2
Representation of the cells Z_i of (C.3) for Example C.2a)

$p_1(x,y) = -y^2 + x^2 + 3x$
$p_2(x,y) = y^2 + 0,3x^2 - 5$

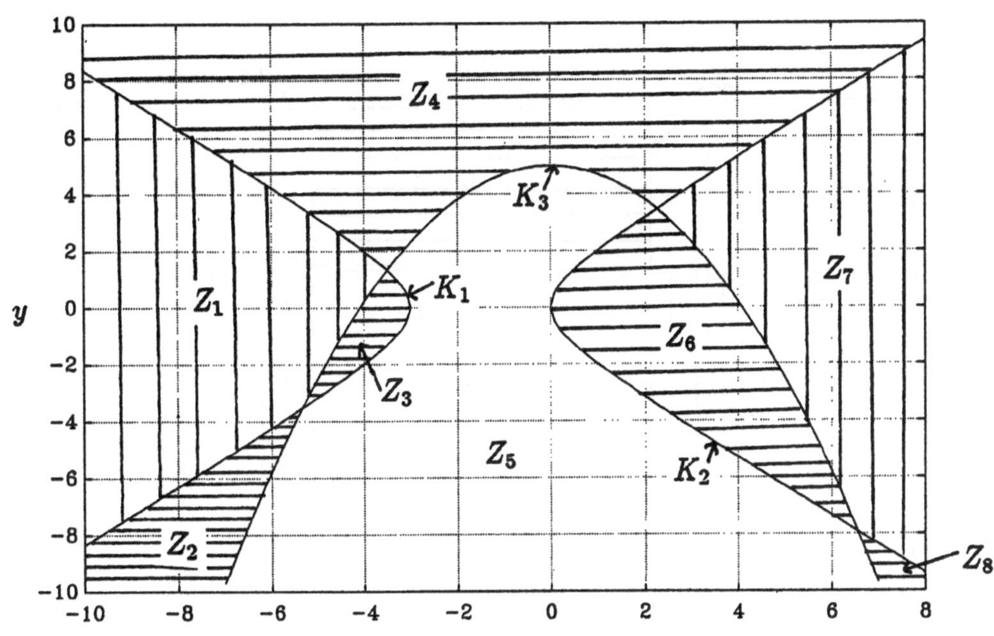

$K_1 = \{\binom{x}{y} | y^2 = x^2 + 3x, x \leq -3\}$
$K_2 = \{\binom{x}{y} | y^2 = x^2 + 3x, x \geq 0\}$
$K_3 = \{\binom{x}{y} | y = 5 - 0,3x^2\}$
$Z_1 = \{\binom{x}{y} | y^2 < x^2 + 3x, y > 5 - 0,3x^2, x < -3\}$
$Z_2 = \{\binom{x}{y} | y^2 > x^2 + 3x, y > 5 - 0,3x^2, x < 0, y < 0\}$
$Z_3 = \{\binom{x}{y} | y^2 < x^2 + 3x, y < 5 - 0,3x^2, x < -3\}$
$Z_4 = \{\binom{x}{y} | y^2 > x^2 + 3x, y > 5 - 0,3x^2, y > 0\}$
$Z_5 = \{\binom{x}{y} | y^2 > x^2 + 3x, y < 5 - 0,3x^2\}$
$Z_6 = \{\binom{x}{y} | y^2 < x^2 + 3x, y < 5 - 0,3x^2, x > 0\}$
$Z_7 = \{\binom{x}{y} | y^2 < x^2 + 3x, y > 5 - 0,3x^2, x > 0\}$
$Z_8 = \{\binom{x}{y} | y^2 > x^2 + 3x, y > 5 - 0,3x^2, x > 0, y < 0\}$

Legend: Z_i - i-th cell
K_i - i-th curve of the point set K

Fig. C.3
Representation of the cells[182] Z_i of (C.3) for Example C.2b)

[182] Cf. the cells in Fig. C.1.b).

$p_1(x,y) = x^2 - 16$
$p_1(x,y) = y^2 - 16$

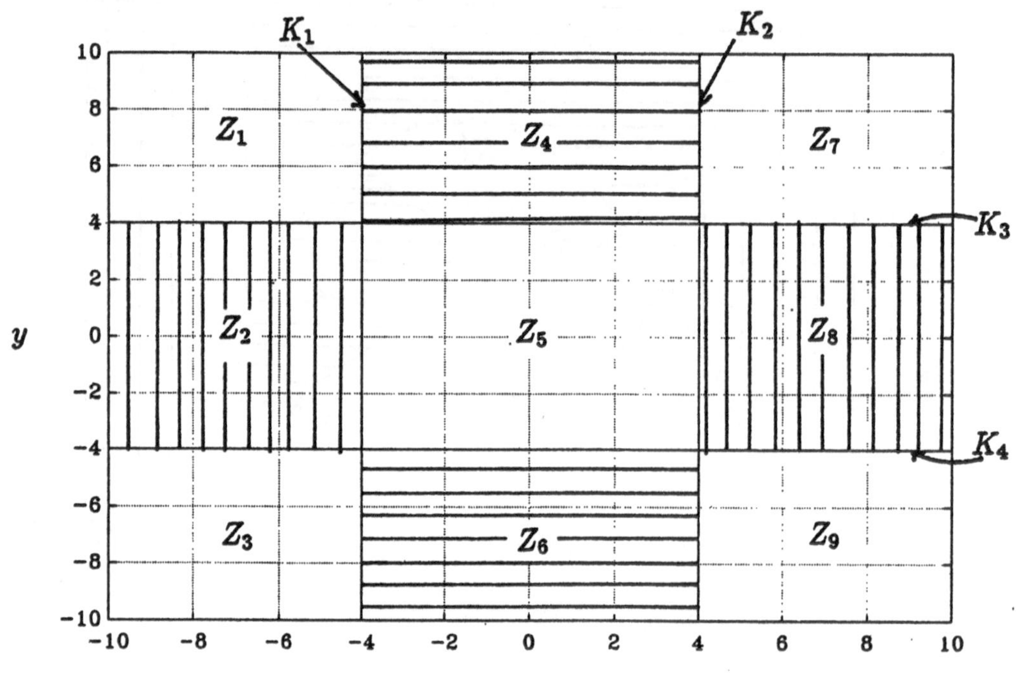

$Z_1 = \{\binom{x}{y} | x < -4, y > 4\}$
$Z_2 = \{\binom{x}{y} | x < -4, |y| < 4\}$
$Z_3 = \{\binom{x}{y} | x < -4, y < -4\}$
$Z_4 = \{\binom{x}{y} | |x| < 4, y > 4\}$
$Z_5 = \{\binom{x}{y} | |x| < 4, |y| < 4\}$
$Z_6 = \{\binom{x}{y} | |x| < 4, y < -4\}$
$Z_7 = \{\binom{x}{y} | x > 4, y > 4\}$
$Z_8 = \{\binom{x}{y} | x > 4, |y| < 4\}$
$Z_9 = \{\binom{x}{y} | x > 4, y < -4\}$

$K_1 = \{\binom{x}{y} | x = -4\}$
$K_2 = \{\binom{x}{y} | x = 4\}$
$K_3 = \{\binom{x}{y} | y = 4\}$
$K_4 = \{\binom{x}{y} | y = -4\}$

Legend: Z_i - i-th cell
K_i - i-th curve of the point set K

Fig. C.4
Representation of the cells Z_i of (C.3) for Example C.2c)

(cf. Examples C.1.b), C.2.b)), the three cells in (C.4) are further subdivided by the "new" curve K_3 (cf. Fig. C.1.b) and Fig. C.3). Thus eight "new" cells result altogether.

Hence we derive a rough estimation for the maximal cell number of a nonlinear inequality system as follows:

For a single inequality of the form (C.2), we obtain one or two curves K_i, which "normally" subdivide the space $\mathbb{R}^2$ in maximal three cells (cf. Fig. C.1). An additional inequality of the form (C.2) yields at most two further curves K_i, subdividing each existing cell in at most three parts. This results in maximal $3^2 = 9$ "new" cells (cf. Fig. C.4). For a system of n inequalities of the form (C.2) we obtain at most 3^n cells. Finally

$$(k+1)^n$$

is an estimation of the maximal cell number if all polynomials in the last-named system have degree k.

This consideration yields

$$4^{10} \cdot 3^{12} \cdot 2^7 \approx 7,13 \cdot 10^{13} \tag{C.6}$$

as an estimation of the maximal cell number of system (C.1). The union of all these cells represents a partition of the space $\mathbb{R}^{12}$ (cylindrical algebraic decomposition (CAD)[183]).

In order to solve (C.1), the cells have to be determined for which all p_i are positive. There exist diverse algorithms for determining a CAD or certain subsystems of cells[184], but the mere construction of a CAD with "only" 489 cells required about 18 hours[185]. Thus the above algorithms are not suitable for solving determinant inequality systems of the form (C.1) yielding "astronomical" cell numbers (cf. (C.6)).

---

183 Cf. e.g. Arnon et al. (1984).

184 Cf. Arnon (1985a, b), Arnon et al. (1984, 1985) and the literature cited there.

185 Cf. Arnon (1985b: 270).

References

Abel, P. (1987).
"On the choice of the pivot columns of the simplex-method: Gradient criteria." *Computing* 38: 13-21.

Adler, I., Karp, R. & Shamir, R. (1986).
"A family of simplex variants solving an $m \times d$ linear program in expected number of pivot steps depending on d only." *Mathematics of Operations Research*, 11, 4: 570-590.

Ahrens, J.H. & Finke, G. (1975).
"Degeneracy in fixed cost transportation problems". In: *Mathematical Programming* 8: 369-374.

Ahuja, R.K. (1985).
"Minimax linear programming problem". *Operations Research Letters* 4, 1: 131-134.

Aigner, M. (1975).
Kombinatorik I: Grundlagen und Zähltheorie. Berlin, Heidelberg, New-York.

Akgül, M. (1987).
A genuinely polynomial primal simplex algorithm for the assignment problem. SERC Report IEOR 87-07: Bilkent University.

Altherr, W. (1975).
"An algorithm for enumerating all vertices of a convex polyhedron". *Computing* 15, 181-193.

Altman, M. (1964).
"Optimum simplex methods and degeneracy in linear programming". *Bulletin de l'Academie Polonaise des Sciences* 12, 217-225.

Arnon, D.S. (1985a).
"A cluster-based cylindrical algebraic decomposition algorithm". In: Buchberger. B. (Ed.). Proceedings of EUROCAL '85. *Lecture Notes in Computing Science* No. 204. (pp. 262-269).

Arnon, D.S. (1985b).
"On mechanical quantifier elimination for elementary algebra and geometry: solution of a nontrivial problem". In: Buchberger. B. (Ed.). Proceedings of EUROCAL '85. *Lecture Notes in Computing Science* No. 204. (pp. 270-271).

Arnon, D.S., Collins, G.E. & McCallum, S. (1984).
"Cylindrical algebraic decomposition I: the basic algorithm". In: *SIAM J. Comput.* 13, 865-877.

Arnon, D.S., Collins, G.E. & McCallum, S. (1985).
"An adjacency algorithm for cylindrical algebraic decomposition of three-dimensional space". In: Buchberger. B. (Ed.). Proceedings of EUROCAL '85. *Lecture Notes in Computing Science* No. 204. (pp. 246-261).

Avis, D. & Chvátal, V. (1978).
"Notes on BLAND's pivoting rule". *Mathematical Programming Study* 8: 24-34.

Azpeitia, A.G. & Dickinson, D.J. (1964).
"A decision rule in the simplex method that avoids cycling". *Numerische Mathematik* 6: 329-331.

Bachem, A. (1980).
"Komplexitätstheorie im Operations Research". *Zeitschrift für Betriebswirtschaft*, 7: 812-844.

Balas, E. (1971).
"Intersection cuts – a new type of cutting planes for integer programming". *Operations Research* 19: 19-39.

Balinsky, M.L., Liebling, Th.M. & Nobs, A.-E. (1986).
"On the average length of lexicographic paths". *Mathematical Programming* 35, 362-364.

Barr, R.S., Glover, F. & Klingman, D. (1977).
"The alternating basis algorithm for assignment problems". *Mathematical Programming* 13: 1-13.

Barr, R.S., Glover, F. & Klingman, D. (1978).
"The generalized alternating path algorithm for transportation problems". *European Journal of Operations Research* 2, 137-144.

Beale, E.M.L. (1955).
"Cycling in the dual simplex algorithm". *Naval Research Logistics Quarterly* 2, 269-276.

Behzad, M. & Chartrand, G. (1971).
Introduction to the theory of graphs. Boston.

Beineke, L.W. & Wilson, R.J. (1978).
Selected topics in graph theory. London, New-York, San Francisco.

Beisel, E.-P. & Mendel, M. (1987).
Optimierungsmethoden des Operations Research,Band 1. Braunschweig, Wiesbaden.

Benichou, M., Gauthier, J.M., Hentges, G. & Ribiere, G. (1977).
"The efficient solution of large-scale linear programming problems – some algorithmic techniques and computational results". *Mathematical Programming* 13, 280-322.

Bitz, M. (1988).
Entscheidungstheorie im Nebenfach, Kurseinheit 5: Spieltheoretische Ansätze. Fernuniversität Hagen.

Bixby, R.E. & Cunningham, W.H. (1980).
"Converting linear programs to network problems". *Mathematics of Operations Research* 5, 321-357.

Bixby, R.E. & Wagner, D.K. (1987).
"A note on detecting simple redundancies in linear systems". *Operations Research Letters* 6, 1, 15-17.

Bland, R.G. (1977).
"New finite pivoting rules for the simplex method". *Mathematics of Operations Research* 2, 103-107.

Boenchendorf, K. (1987).
"An optimality criterion for degenerated transportation problems". In: Domschke, W. et al. (Ed.). *Methods of Operations Research* 57, XI. Symposium on Operations Research. (pp. 87-94). Frankfurt a.M.

Borgwardt, K.H. (1982).
"The average number of pivot steps required by the simplex method is polynomial". *Zeitschrift für Operations Research* 26, 157-177.

Borgwardt, K.H. (1985).
"Der durchschnittliche Rechenaufwand beim Simplexverfahren". In: Ohse, D. et al. (Ed.). *Operations Research Proceedings* 1984 (pp. 647-660). Berlin, Heidelberg.

Borgwardt, K.H. (1987).
The simplex method – a probabilistic approach. Berlin, Heidelberg.

Buchberger, B. (Ed.) (1985).
Proceedings of EUROCAL '85, *Lecture Notes in Computing Science* No. 204.

Burkard, R.E. (1972).
Methoden der ganzzahligen Optimierung. Wien, New-York.

Burkard, R.E. (1987).
"Ganzzahlige Optimierung". In: Gal, T. (Ed.). Grundlagen des Operations Research, Band 2 (pp. 361ff). Berlin, Heidelberg, New-York.

Calamai, P.H. & Moré, J.J. (1987).
"Projected gradient methods for linearly constrained problems". *Mathematical Programming* 39, 93-116.

Cameron, N. (1987).
"Stationarity in the simplex method". *J. Austral. Math. Soc. (Series A)* 43, 137-142.

Chang, Y.-Y. & Cottle, R.W. (1980).
"Least-index resolution of degeneracy in quadratic programming". *Mathematical Programming* 18, 27-137.

Charnes, A. (1952).
"Optimality and degeneracy in linear programming". *Econometrica* 20, 160-170.

Charnes, A. & Cooper, W.W. (1954).
"The stepping stone method of explaining linear programming calculations in transportation problems". *Management Science* 1, 1.

Cheng, M.C. (1980).
"New criteria for the simplex algorithm". *Mathematical Programming* 19, 230-236.

Cirinà, M. (1985).
"Remarks on a recent simplex pivoting rule". In: Domschke, W. et al. (Ed). *Methods of Operations Research* 55, IX. Symposium on Operations Research. (pp. 187-199). Königsstein/Ts.

Cirinà M. (1989).
Efficient finite pivoting rules for the simplex method. Unpublished paper.

Clausen, J. (1987).
"A new family of exponential LP problems". *European Journal of Operational Research* 32, 130-139.

Cooper, L. & Steinberg, D.I. (1974).
Methods and applications of linear programming. Philadelphia.

Cunningham, W.H. (1976).
"A network simplex method". *Mathematical Programming* 11, 105-116.

Cunningham, W.H. (1979).
"Theoretical properties of the network simplex method". *Mathematics of Operations Research* 4, 196-208.

Cunningham, W.H. & Klincewicz, J.G. (1983).
"On cycling in the network simplex method". *Mathematical Programming* 26, 182-189.

Dantzig, G.B. (1951).
"Maximization of a linear function of variables subject to linear inequalities". In: Koopmans, T.C. (Ed.). Activity analysis of production and allocation. New-York.

Dantzig, G.B. (1951a).
"Application of the simplex method to a transportation problem". In: Koopmans, T.C. (Ed.). Activity analysis of production and allocation. New-York.

Dantzig, G.B. (1966).
Lineare Programmierung und Erweiterungen. Berlin, Heidelberg, New-York.

Dantzig, G.B., Orden, A. & Wolfe, P. (1955).
"The generalized simplex method for minimizing a linear form under linear inequality restraints". *Pacific Journal of Mathematics* 5, 183-195.

Derigs, U. (1986).
"Neuere Ansätze in der Linearen Optimierung – Motivation, Konzepte und Verfahren". In: Streitferdt, L. et al. (Ed.). *Operations Research Proceedings* 1985. (pp. 47-58). Berlin, Heidelberg.

Domschke, W. (1981).
Logistik: Transport. München, Wien.

Dyer, M.E. & Proll, L.G. (1980).
"Eliminating extraneous edges in Greenberg's algorithm". *Mathematical Programming* 19, 106-110.

Dyer, M.E. & Proll, L.G. (1982).
"An improved vertex enumeration algorithm". *European Journal of Operational Research* 9, 359-368.

Ecker, J.G. & Kupferschmid, M. (1988).
Introduction to operations research. New-York, Chichester, Brisbane, Toronto, Singapore.

Edmonds J. (1965).
"Maximum matching and a polyhedron with 0,1 vertices". *J. Res. NBS, 69B*, 125-130.

Evans, J.R. & Baker, N.R. (1982).
"Degeneracy and the (mis-)interpretation of sensitivity analysis in linear programming". *Decision Sciences* 13, 348-354.

Fleischmann, B. (1970).
Duale und primale Schnitthyperebenenverfahren in der ganzzahligen linearen Optimierung. Thesis at University of Hamburg.

Fletcher, R. (1987).
Practical methods of optimization. New-York, Chichester, Brisbane, Toronto, Singapore.

Fourer, R. (1988).
"A simplex algorithm for piecewise-linear programming II: finiteness, feasibility and degeneracy". *Mathematical Programming* 41, 281-315.

Fournier, I. (1985).
"Longest cycles in 2-connected graphs of independence number α". *Annals of Discrete Mathematics* 27, 201-204.

Fraisse, P. (1986).
"Circuits including a given set of vertices." *Journal of Graph Theory*, Vol. 10, 553-557.

Frieze, A.M. (1975).
"Bottleneck linear programming". *Operational Research Quarterly* 26, 871-874.

Gal, T. (1978).
Determination of all neighbours of a degenerate extreme point in polytopes. Discussion paper No. 17b, Fernuniversität Hagen.

Gal, T. (1979).
Postoptimal analyses, paramatric programming and related topics. New York.

Gal, T. (1983).
"A method for determining redundant constraints". In: Karwan, M.H., Lotfi, V., Telgen, J. & Zionts, S. (Ed.). Redundancy in mathematical programming: A state-of-the-art survey. Berlin, Heidelberg, New-York, Tokyo, pp. 36-52.

Gal, T. (1983a).
Mathematik für Wirtschaftswissenschaftler, Band III: Lineare Optimierung. Berlin, Heidelberg, New-York, Tokyo.

Gal, T. (1986).
"Shadow prices and sensitivity analysis in linear programming under degeneracy: A state-of-the-art survey". *Operations Research Spektrum*,8, 59-71.

Gal, T. (Ed.) (1987).
Grundlagen des Operations Research. Berlin, Heidelberg, New York.

Gal, T. (1988).
Degeneracy Graphs – Theory and Application: A state-of-the-art survey. Discussion paper No. 126, Fernuniversität Hagen.

Gal, T. & Kruse, H.-J. (1984).
"Ein Verfahren zur Lösung des Nachbarschaftsproblems". *Operations Research Verfahren*, 447-454.

Gal, T., Kruse, H.-J. & Zörnig, P. (1986).
New developments in the area of degeneracy graphs. Discussion paper No. 102, Fernuniversität Hagen. Presented at Joint National Meeting (TIMS/ORSA), Los Angeles.

Gal, T., Kruse, H.-J. & Zörnig, P. (1988).
"Survey of solved and open problems in the degeneracy phenomenon". *Mathematical Programming* 42, 125-133.

Gale, D. (1969).
"How to solve linear inequalities". *American Mathematical Monthly* 76, 589-599.

Garfinkel, R.S. & Nemhauser, G.L. (1972).
Integer Programming. New-York, London, Sydney, Toronto.

Gass, S.I. (1979).
"Comments on the possibility of cycling with the simplex method". *Operations Research* 27, 848-852.

Gass, S.I. (1985).
Linear Programming: Methods and Applications. 5.Edition, Kingsport.

Gassner, B.J. (1964).
"Cycling in the transportation problem." *Naval Research Logistics Quarterly* 11, 43-58.

Garfinkel, R.S. & Rao, M.R. (1976).
"Bottleneck linear programming". *Mathematical Programming* 11, 291-298.

Geue, F. (1989).
Eine neue Pivotauswahlregel und die durch sie induzierten Teilgraphen des positiven Entartungsgraphen. Discussion paper No. 141, FB WIWI, Fernuniversität Hagen.

Glover, F. (1968).
"A new foundation for a simplified primal integer programming algorithm". *Operations Research* 16, 727-740.

Götz, R. (1989).
Verfahren zur Konstruktion von Beispielen zum Simplexzykeln und ihre Implementierung. Diplomarbeit FB WIWI, Fernuniversität Hagen.

Goldfarb, D. & Sit, W.Y. (1979).
"Worst case behaviour of the steepest edge simplex method". *Discrete Applied Mathematics* 1, 277-285.

Gomory, R.E. (1963).
"An all-integer programming algorithm". In: Muth, J.F. & Thompson, G.L. (Eds.). Industrial scheduling. (pp. 193-206). New-York.

Gotterbarm, F. (1983).
"Zur Auswahl der Pivotspalte im Simplex-Algorithmus". In: Bühler, W. et al. (Eds.). *Operations Research Proceedings* 1982. (pp. 566-576). Berlin, Heidelberg.

Greenberg, H.J. (1986).
"An analysis of degeneracy". *Naval Research Logistics Quarterly* 33, 635-655.

Grünbaum, B. (1967).
Convex polytopes. London, New York.

Gupta, H. (1962).
Tables of partitions. Royal Soc. Math. Tables, Vol. 4, Cambridge University Press.

Hadley, G. (1975).
Linear programming. 9th printing. London, Massachusetts.

Harary, F. (1974).
Graphentheorie. München, Wien.

Hardy, G.H. (1959).
Ramanujan. London 1940. (Repr. New York 1959: Chelsea).

Hardy, G.H. & Wright, E.M. (1958).
Einführung in die Zahlentheorie. München.

Harris, P.M.J. (1973). "Pivot selection methods of the DEVEX LP Code". *Mathematical Programming* 5, 1-28.

Hattersley, B. & Wilson, J. (1988).
"A dual approach to primal degeneracy". *Mathematical Programming* 42, 135-145.

Hoffmann, A.J. (1953).
Cycling in the simplex algorithm. National Bureau of Standards Report 2974.

Horst, R. (1987).
"Nichtlineare Optimierung". In: Gal, T. (Ed.). Grundlagen des Operations Research. Band 1, pp. 255ff. Berlin, Heidelberg, New York.

Horst, R., de Vries, J. & Thoai, N.V. (1988).
"On finding new vertices and redundant constraints in cutting plane algorithms for global optimization". *Operations Research Letters* 7,2,85-90.

Jaeger, F. (1985).
"A survey of the cycle double cover conjecture". *Annals of Discrete Mathematics* 27, 1-12.

Jansson, Ch. (1985).
Zur linearen Optimierung mit unscharfen Daten. Thesis at University of Kaiserslautern.

Jeroslow, R.G. (1973).
"The simplex algorithm with the pivot rule of maximizing criterion improvement". *Discrete Mathematics* 4, 367-377.

Joksch, H.C. (1965).
Lineares Programmieren. Tübingen, 2. Edition.

Karmarkar, N. (1984).
"A new polynomial time algorithm for Linear Programming". *Combinatorica* 4, 373-395.

Karwan, M.H., Lotfi, V., Telgen, J. & Zionts, S. (Eds.) (1983).
Redundancy in mathematical programming: a state-of-the-art survey. Berlin, Heidelberg, New York, Tokyo.

Keller, E.L. (1973).
"The general quadratic optimization problem". *Mathematical Programming* 5, 311-337.

Khachian, L.G. (1979).
"A polynomial algorithm in linear programming". *Doklady Akademia Nauk SSSR* 244, 1093-1096.

Kiehne, R. (1969).
Innerbetriebliche Standortplanung und Raumzuordnung. Wiesbaden.

Klee, V. & Minty, G.J. (1972).
"How good is the simplex algorithm?" In: Shisha, O. (Ed.). Inequalities III. New York, London.

Knolmayer, G. (1984).
"The effects of degeneracy on cost-coefficient ranges and an algorithm to resolve interpretation problems". *Decision Sciences* 15, 14-21.

Kostreva, M.M. (1979).
"Cycling in linear complementarity problems". *Mathematical Programming* 16, 127-130.

Kotiah, T.C.T. & Steinberg, D.I. (1977). "Occurrences of cycling and other phenomena arising in a class of linear programming models". *Communications of the Association for Computing Machinery* 20, 107-112.

Kotiah, T.C.T. & Steinberg, D.I. (1978). "On the possibility of cycling with the simplex method". *Operations Research* 26, 374-376.

Kowalsky, H.-J. (1975).
Lineare Algebra. Berlin, New York.

Kruse, H.-J. (1984).
Entartungsgraphen und ihre Anwendung zur Lösung des Nachbarschaftsproblems. Thesis at Fernuniversität Hagen.

Kruse, H.-J. (1984a).
Zur Theorie und Anwendung von Entartungsgraphen. Discussion paper No. 86, FB WIWI, Fernuniversität Hagen.

Kruse, H.-J. (1986).
Degeneracy graphs and the neighbourhood problem. Lecture Notes in Economics and Mathematical Systems 260. Berlin, Heidelberg, New York, Tokyo.

Kruse, H.-J. (1987).
Über spezielle Teilgraphen von Entartungsgraphen. Discussion paper No. 121, FB WIWI, Fernuniversität Hagen.

Kuhn, H.W. & Quandt, R.E. (1963).
"An experimental study of the simplex method". *Proc. Symposia in Appl. Math.* 15, 107-124.

Lemke, C.E. (1965).
"Bimatrix equilibrium points and mathematical programming". *Management Science* 11, 681-689.

Liebling, Th.M. (1973).
"On the number of iterations of the simplex method". *Operations Research Verfahren* 17, 248-264.

Madan Lal Mittal (1967).
"A note on resolution of degeneracy in transportation problems". *Operational Research Quarterly* 19, 175-184.

Magnanti, T.L. & Orlin, J.B. (1988).
"Parametric linear programming and anti-cycling pivoting rules". *Mathematical Programming* 41, 317-325.

Majthay, A. (1981).
On degeneracy and cycling with the simplex method. Discussion paper No. 41, Center for Econometrics and Decision Sciences, University of Florida.

Marshall, K.T. & Suurballe, J.W. (1969).
"A note on cycling in the simplex method". *Naval Research Logistics Quarterly* 16, 121-137.

Mathies, S. (1989).
Entartungsprobleme bei der Lösung von Optimierungsaufgaben. Diplomarbeit, FB WIWI, Fernuniversität Hagen.

McKeown, P.G. (1978).
"Some computational results of using the Ahrens-Finke method for handling degeneracy in fixed charge transportation problems". *Mathematical Programming* 15, 355-359.

Megiddo, N. (1986).
"A note on degeneracy in linear programming". *Mathematical Programming* 35, 365-367.

Megiddo, N. (1986a).
"Improved asymptotic analysis of the average number of steps performed by the self-dual simplex algorithm". *Mathematical Programming* 35, 140-172.

Mlynarovic, V. (1988).
"On shadow prices in convex programming". *Econ. Mat. Obzor.* 29, 201-214.

Müller-Merbach, H. (1973).
Operations Research. 3. Edition. Vahlen, München.

Nash, P. (1985).
"Algebraic fundamentals of linear programming". *Lect. Notes Econ. Math. Syst.* 259, 37-52.

Nelson, R.R. (1957).
"Degeneracy in linear programming: A simple geometric interpretation". *Review of Economics and Statistics* 39, 402-407.

Nygreen, B. (1987).
"A possible way to reduce degeneracy in integer programming computations". *Operations Research Letters* 6, 47-51.

Ohse, D. (1987).
Transportprobleme. In:Gal, T. (Ed.). Grundlagen des Operations Research. Band 2. Berlin, Heidelberg, New York. p.261ff.

Ollmert, H.-J. (1965).
Zur Theorie des Simplex-Verfahrens im ausgearteten Fall. Diplomarbeit, University of Saarbrücken.

Ollmert, H.-J. (1969).
"Kreisende lineare Programme". In: Henn, R., Künzi, H.P., & Schubert, H. (Eds.). Operations Research Verfahren VI. Meisenheim am Glan. (pp. 186-216).

Perold. A.F. (1980).
"A degeneracy exploiting LU factorization for the simplex method." *Mathematical Programming* 19, 239-254.

Philip, J. (1977).
"Vector maximization at a degenerate vertex". *Mathematical Programming* 13, 357-359.

Piehler, G. (1988).
Optimalbasenansatz zur Sensitivitätsanalyse bei linearer Programmierung unter Entartung. Discussion paper No. 130, FB WIWI, Fernuniversität Hagen.

Piehler, G. & Kruse, H.-J. (1989).
Optimumgraphen zur Analyse linearer Optimierungsprobleme unter Entartung. *Operations Research Proceedings* 1988. Berlin, Heidelberg.

Proll, L.G. (1987).
"Goal aggregation via shadow prices - Some counterexamples". *Large Scale Systems* 12, 83-85.

Ramesh, R., Karwan, M.H. & Zionts, S. (1987).
"Degeneracy in efficiency testing in bi-criteria integer programming". Presented at *Methodology and Software for Interactive Decision Support*, IIASA Laxenburg, 1987.

Richards, D., & Liestman, A.L. (1985).
"Finding cycles of a given length". *Annals of Discrete Mathematics* 27, 249-256.

Ritter, K. (1984).
"On parametric linear and quadratic programming problems". *Mathematical Programming* 92, 307-335.

Roos, C. (1984).
"The umbrella approach to linear programming". *Proceedings der Jahrestagung Mathematische Optimierung* 16, 67-79.

Ruszczynski, A. (1986).
"A regularized decomposition method for minimizing a sum of polyhedral functions". *Mathematical Programming* 35, 309-333.

Ryan, D.M. & Osborne, M.R. (1988).
"On the solution of highly degenerate linear programmes". *Mathematical Programming* 41, 385-392.

Schwödiauer, G. (1987).
"Spieltheorie". In Gal, T. (Ed.). Grundlagen des Operations Research. Band 3, pp. 1 ff., Berlin, Heidelberg, New York.

Seshan, C.R. & Achary, K.K. (1982).
"On the bottleneck linear programming problem". *European Journal of Operational Research* 9, 347-352.

Shamir, R. (1987).
"The efficiency of the simplex method: A survey". *Management Science* 33,3,301-334.

Sherali, H.D. & Dickey, S.E. (1986).
"An extreme-point-ranking algorithm for the extreme-point mathematical programming problem". *Comput. & Ops. Res.* 13,4, 465-475.

Smale, S. (1983).
"On the average number of steps of the simplex method of linear programming". *Mathematical Programming* 27, 241-262.

Solow, D. (1984).
Linear programming. An introduction to finite improvement algorithms. New York, Amsterdam, Oxford.

Telgen, J. (1980).
A note on a linear programming problem that cycled". *COAL Newsletter* 2, 8-11.

Telgen, J. (1983).
"Identifying redundancy in systems of linear constraints". In: Karwan, M.H., Lotfi, V., Telgen, J. & Zionts, S. (Eds.) Redundancy in mathematical programming: a state-of-the-art survery. (pp. 53-59). Berlin, Heidelberg, New York, Tokyo.

Thompson, G.L., Tonge, F.M. & Zionts, S. (1966).
"Techniques for removing nonbinding constraints and extraneous variables from linear programming problems". *Management Science* 12,7, 588-608.

Todd, M.J. (1982).
"An implementation of the simplex method for linear programming problems with variable upper bounds". *Mathematical Programming* 23, 34-49.

Todd, M.J. (1986).
"Polynomial expected behavior of a pivoting algorithm for linear complementarity and linear programming problems". *Mathematical Programming* 35, 173-192.

Tomlin, J.A. (1970).
"Branch and bound methods for integer and non-convex programming". In: Abadie, J. (Ed.). Integer and Nonlinear Programming. (pp. 437-450). Amsterdam.

Tomlin, J.A. & Welch, J.S. (1983).
"Formal optimization of some reduced linear programming problems". *Mathematical Programming* 27, 232-240.

Varga, J. (1974).
Praktische Optimierung. München, Wien.

Vöros, J. (1987).
"The explicit derivation of the efficient portfolio frontier in the case of degeneracy and general singularity". *European Journal of Operational Research* 32, 302-310.

Vogel, W. (1967).
Lineares Optimieren. Leipzig.

Volgenant, A., Jonker, R. & Kindervater, G.A.P. (1986).
"A note on finding a shortest complete cycle in an undirected graph". *European Journal of Operational Research* 23, 82-85.

Wagner, K. (1970).
Graphentheorie. Mannheim, Wien, Zürich.

Wallace, S.W. (1985).
"Pivoting rules and redundancy schemes in extreme point enumeration". *BIT* 25, 274-280.

Wolfe, P. (1963).
"A technique for resolving degeneracy in linear programming". *Journal of the Society for Industrial and Applied Mathematics* 11, 305-311.

Young, R.D. (1968).
"A simplified primal (all-integer) integer programming algorithm". *Operations Research* 16, 750-782.

Yudin, D.B. & Gol'shtein, E.G: (1965).
Linear Programming. Israel Program for Scientific Translations. Jerusalem.

Zadeh, N. (1973).
"A bad network problem for the simplex method and other minimum cost flow algorithms". *Mathematical Programming* 5, 255-266.

Zimmermann, H.-J. & Gal, T. (1975).
"Redundanz und ihre Bedeutung für betriebliche Optimierungsentscheidungen": *Zeitschrift für Betriebswirtschaft* 45, 221-236.

Zörnig, P. (1985).
Strukturuntersuchungen an 2 × n-Entartungsgraphen. Discussion paper No. 87, FB WIWI, Fernuniversität Hagen.

Zörnig, P. (1990).
Theorie der Entartungsgraphen und ihre Anwendung zur Erklärung des Simplexzykelns. Thesis at Fernuniversität Hagen.

Vol. 261: Th. R. Gulledge, Jr., N. K. Womer, The Economics of Made-to-Order Production. VI, 134 pages. 1986.

Vol. 262: H. U. Buhl, A Neo-Classical Theory of Distribution and Wealth. V, 146 pages. 1986.

Vol. 263: M. Schäfer, Resource Extraction and Market Structure. XI, 154 pages. 1986.

Vol. 264: Models of Economic Dynamics. Proceedings, 1983. Edited by H.F. Sonnenschein. VII, 212 pages. 1986.

Vol. 265: Dynamic Games and Applications in Economics. Edited by T. Başar. IX, 288 pages. 1986.

Vol. 266: Multi-Stage Production Planning and Inventory Control. Edited by S. Axsäter, Ch. Schneeweiss and E. Silver. V, 264 pages. 1986.

Vol. 267: R. Bemelmans, The Capacity Aspect of Inventories. IX, 165 pages. 1986.

Vol. 268: V. Firchau, Information Evaluation in Capital Markets. VII, 103 pages. 1986.

Vol. 269: A. Borglin, H. Keiding, Optimality in Infinite Horizon Economies. VI, 180 pages. 1986.

Vol. 270: Technological Change, Employment and Spatial Dynamics. Proceedings 1985. Edited by P. Nijkamp. VII, 466 pages. 1986.

Vol. 271: C. Hildreth, The Cowles Commission in Chicago, 1939–1955. V, 176 pages. 1986.

Vol. 272: G. Clemenz, Credit Markets with Asymmetric Information. VIII, 212 pages. 1986.

Vol. 273: Large-Scale Modelling and Interactive Decision Analysis. Proceedings, 1985. Edited by G. Fandel, M. Grauer, A. Kurzhanski and A. P. Wierzbicki. VII, 363 pages. 1986.

Vol. 274: W.K. Klein Haneveld, Duality in Stochastic Linear and Dynamic Programming. VII, 295 pages. 1986.

Vol. 275: Competition, Instability, and Nonlinear Cycles. Proceedings, 1985. Edited by W. Semmler. XII, 340 pages. 1986.

Vol. 276: M.R. Baye, D.A. Black, Consumer Behavior, Cost of Living Measures, and the Income Tax. VII, 119 pages. 1986.

Vol. 277: Studies in Austrian Capital Theory, Investment and Time. Edited by M. Faber. VI, 317 pages. 1986.

Vol. 278: W. E. Diewert, The Measurement of the Economic Benefits of Infrastructure Services. V, 202 pages. 1986.

Vol. 279: H.-J. Büttler, G. Frei and B. Schips, Estimation of Disequilibrium Models. VI, 114 pages. 1986.

Vol. 280: H.T. Lau, Combinatorial Heuristic Algorithms with FORTRAN. VII, 126 pages. 1986.

Vol. 281: Ch.-L. Hwang, M.-J. Lin, Group Decision Making under Multiple Criteria. XI, 400 pages. 1987.

Vol. 282: K. Schittkowski, More Test Examples for Nonlinear Programming Codes. V, 261 pages. 1987.

Vol. 283: G. Gabisch, H.-W. Lorenz, Business Cycle Theory. VII, 229 pages. 1987.

Vol. 284: H. Lütkepohl, Forecasting Aggregated Vector ARMA Processes. X, 323 pages. 1987.

Vol. 285: Toward Interactive and Intelligent Decision Support Systems. Volume 1. Proceedings, 1986. Edited by Y. Sawaragi, K. Inoue and H. Nakayama. XII, 445 pages. 1987.

Vol. 286: Toward Interactive and Intelligent Decision Support Systems. Volume 2. Proceedings, 1986. Edited by Y. Sawaragi, K. Inoue and H. Nakayama. XII, 450 pages. 1987.

Vol. 287: Dynamical Systems. Proceedings, 1985. Edited by A.B. Kurzhanski and K. Sigmund. VI, 215 pages. 1987.

Vol. 288: G.D. Rudebusch, The Estimation of Macroeconomic Disequilibrium Models with Regime Classification Information. VII, 128 pages. 1987.

Vol. 289: B.R. Meijboom, Planning in Decentralized Firms. X, 168 pages. 1987.

Vol. 290: D.A. Carlson, A. Haurie, Infinite Horizon Optimal Control. XI, 254 pages. 1987.

Vol. 291: N. Takahashi, Design of Adaptive Organizations. VI, 140 pages. 1987.

Vol. 292: I. Tchijov, L. Tomaszewicz (Eds.), Input-Output Modeling. Proceedings, 1985. VI, 195 pages. 1987.

Vol. 293: D. Batten, J. Casti, B. Johansson (Eds.), Economic Evolution and Structural Adjustment. Proceedings, 1985. VI, 382 pages. 1987.

Vol. 294: J. Jahn, W. Krabs (Eds.), Recent Advances and Historical Development of Vector Optimization. VII, 405 pages. 1987.

Vol. 295: H. Meister, The Purification Problem for Constrained Games with Incomplete Information. X, 127 pages. 1987.

Vol. 296: A. Börsch-Supan, Econometric Analysis of Discrete Choice. VIII, 211 pages. 1987.

Vol. 297: V. Fedorov, H. Läuter (Eds.), Model-Oriented Data Analysis. Proceedings, 1987. VI, 239 pages. 1988.

Vol. 298: S.H. Chew, Q. Zheng, Integral Global Optimization. VII, 179 pages. 1988.

Vol. 299: K. Marti, Descent Directions and Efficient Solutions in Discretely Distributed Stochastic Programs. XIV, 178 pages. 1988.

Vol. 300: U. Derigs, Programming in Networks and Graphs. XI, 315 pages. 1988.

Vol. 301: J. Kacprzyk, M. Roubens (Eds.), Non-Conventional Preference Relations in Decision Making. VII, 155 pages. 1988.

Vol. 302: H.A. Eiselt, G. Pederzoli (Eds.), Advances in Optimization and Control. Proceedings, 1986. VIII, 372 pages. 1988.

Vol. 303: F.X. Diebold, Empirical Modeling of Exchange Rate Dynamics. VII, 143 pages. 1988.

Vol. 304: A. Kurzhanski, K. Neumann, D. Pallaschke (Eds.), Optimization, Parallel Processing and Applications. Proceedings, 1987. VI, 292 pages. 1988.

Vol. 305: G.-J.C.Th. van Schijndel, Dynamic Firm and Investor Behaviour under Progressive Personal Taxation. X, 215 pages. 1988.

Vol. 306: Ch. Klein, A Static Microeconomic Model of Pure Competition. VIII, 139 pages. 1988.

Vol. 307: T.K. Dijkstra (Ed.), On Model Uncertainty and its Statistical Implications. VII, 138 pages. 1988.

Vol. 308: J.R. Daduna, A. Wren (Eds.), Computer-Aided Transit Scheduling. VIII, 339 pages. 1988.

Vol. 309: G. Ricci, K. Velupillai (Eds.), Growth Cycles and Multisectoral Economics: the Goodwin Tradition. III, 126 pages. 1988.

Vol. 310: J. Kacprzyk, M. Fedrizzi (Eds.), Combining Fuzzy Imprecision with Probabilistic Uncertainty in Decision Making. IX, 399 pages. 1988.

Vol. 311: R. Färe, Fundamentals of Production Theory. IX, 163 pages. 1988.

Vol. 312: J. Krishnakumar, Estimation of Simultaneous Equation Models with Error Components Structure. X, 357 pages. 1988.

Vol. 313: W. Jammernegg, Sequential Binary Investment Decisions. VI, 156 pages. 1988.

Vol. 314: R. Tietz, W. Albers, R. Selten (Eds.), Bounded Rational Behavior in Experimental Games and Markets. VI, 368 pages. 1988.

Vol. 315: I. Orishimo, G.J.D. Hewings, P. Nijkamp (Eds.), Information Technology: Social and Spatial Perspectives. Proceedings, 1986. VI, 268 pages. 1988.

Vol. 316: R.L. Basmann, D.J. Slottje, K. Hayes, J.D. Johnson, D.J. Molina, The Generalized Fechner-Thurstone Direct Utility Function and Some of its Uses. VIII, 159 pages. 1988.

Vol. 317: L. Bianco, A. La Bella (Eds.), Freight Transport Planning and Logistics. Proceedings, 1987. X, 568 pages. 1988.

Vol. 318: T. Doup, Simplicial Algorithms on the Simplotope. VIII, 262 pages. 1988.

Vol. 319: D.T. Luc, Theory of Vector Optimization. VIII, 173 pages. 1989.

Vol. 320: D. van der Wijst, Financial Structure in Small Business. VII, 181 pages. 1989.

Vol. 321: M. Di Matteo, R.M. Goodwin, A. Vercelli (Eds.), Technological and Social Factors in Long Term Fluctuations. Proceedings. IX, 442 pages. 1989.

Vol. 322: T. Kollintzas (Ed.), The Rational Expectations Equilibrium Inventory Model. XI, 269 pages. 1989.

Vol. 323: M.B.M. de Koster, Capacity Oriented Analysis and Design of Production Systems. XII, 245 pages. 1989.

Vol. 324: I.M. Bomze, B.M. Pötscher, Game Theoretical Foundations of Evolutionary Stability. VI, 145 pages. 1989.

Vol. 325: P. Ferri, E. Greenberg, The Labor Market and Business Cycle Theories. X, 183 pages. 1989.

Vol. 326: Ch. Sauer, Alternative Theories of Output, Unemployment, and Inflation in Germany: 1960–1985. XIII, 206 pages. 1989.

Vol. 327: M. Tawada, Production Structure and International Trade. V, 132 pages. 1989.

Vol. 328: W. Güth, B. Kalkofen, Unique Solutions for Strategic Games. VII, 200 pages. 1989.

Vol. 329: G. Tillmann, Equity, Incentives, and Taxation. VI, 132 pages. 1989.

Vol. 330: P.M. Kort, Optimal Dynamic Investment Policies of a Value Maximizing Firm. VII, 185 pages. 1989.

Vol. 331: A. Lewandowski, A.P. Wierzbicki (Eds.), Aspiration Based Decision Support Systems. X, 400 pages. 1989.

Vol. 332: T.R. Gulledge, Jr., L.A. Litteral (Eds.), Cost Analysis Applications of Economics and Operations Research. Proceedings. VII, 422 pages. 1989.

Vol. 333: N. Dellaert, Production to Order. VII, 158 pages. 1989.

Vol. 334: H.-W. Lorenz, Nonlinear Dynamical Economics and Chaotic Motion. XI, 248 pages. 1989.

Vol. 335: A. G. Lockett, G. Islei (Eds.), Improving Decision Making in Organisations. Proceedings. IX, 606 pages. 1989.

Vol. 336: T. Puu, Nonlinear Economic Dynamics. VII, 119 pages. 1989.

Vol. 337: A. Lewandowski, I. Stanchev (Eds.), Methodology and Software for Interactive Decision Support. VIII, 309 pages. 1989.

Vol. 338: J.K. Ho, R.P. Sundarraj, DECOMP: an Implementation of Dantzig-Wolfe Decomposition for Linear Programming. VI, 206 pages. 1989.

Vol. 339: J. Terceiro Lomba, Estimation of Dynamic Econometric Models with Errors in Variables. VIII, 116 pages. 1990.

Vol. 340: T. Vasko, R. Ayres, L. Fontvieille (Eds.), Life Cycles and Long Waves. XIV, 293 pages. 1990.

Vol. 341: G.R. Uhlich, Descriptive Theories of Bargaining. IX, 165 pages. 1990.

Vol. 342: K. Okuguchi, F. Szidarovszky, The Theory of Oligopoly with Multi-Product Firms. V, 167 pages. 1990.

Vol. 343: C. Chiarella, The Elements of a Nonlinear Theory of Economic Dynamics. IX, 149 pages. 1990.

Vol. 344: K. Neumann, Stochastic Project Networks. XI, 237 pages. 1990.

Vol. 345: A. Cambini, E. Castagnoli, L. Martein, P. Mazzoleni, S. Schaible (Eds.), Generalized Convexity and Fractional Programming with Economic Applications. Proceedings, 1988. VII, 361 pages. 1990.

Vol. 346: R. von Randow (Ed.), Integer Programming and Related Areas. A Classified Bibliography 1984–1987. XIII, 514 pages. 1990.

Vol. 347: D. Ríos Insua, Sensitivity Analysis in Multi-objective Decision Making. XI, 193 pages. 1990.

Vol. 348: H. Störmer, Binary Functions and their Applications. VIII, 151 pages. 1990.

Vol. 349: G.A. Pfann, Dynamic Modelling of Stochastic Demand for Manufacturing Employment. VI, 158 pages. 1990.

Vol. 350: W.-B. Zhang, Economic Dynamics. X, 232 pages. 1990.

Vol. 351: A. Lewandowski, V. Volkovich (Eds.), Multiobjective Problems of Mathematical Programming. Proceedings, 1988. VII, 315 pages. 1991.

Vol. 352: O. van Hilten, Optimal Firm Behaviour in the Context of Technological Progress and a Business Cycle. XII, 229 pages. 1991.

Vol. 353: G. Ricci (Ed.), Decision Processes in Economics. Proceedings, 1989. III, 209 pages. 1991.

Vol. 354: M. Ivaldi, A Structural Analysis of Expectation Formation. XII, 230 pages. 1991.

Vol. 355: M. Salomon, Deterministic Lotsizing Models for Production Planning. VII, 158 pages. 1991.

Vol. 356: P. Korhonen, A. Lewandowski, J. Wallenius (Eds.), Multiple Criteria Decision Support. Proceedings, 1989. XII, 393 pages. 1991.

Vol. 357: P. Zörnig, Degeneracy Graphs and Simplex Cycling. XV, 194 pages. 1991.

Lecture Notes in Economics and Mathematical Systems

For information about Vols. 1–210, please contact your bookseller or Springer-Verlag

Vol. 211: P. van den Heuvel, The Stability of a Macroeconomic System with Quantity Constraints. VII, 169 pages. 1983.

Vol. 212: R. Sato and T. Nôno, Invariance Principles and the Structure of Technology. V, 94 pages. 1983.

Vol. 213: Aspiration Levels in Bargaining and Economic Decision Making. Proceedings, 1982. Edited by R. Tietz. VIII, 406 pages. 1983.

Vol. 214: M. Faber, H. Niemes und G. Stephan, Entropie, Umweltschutz und Rohstoffverbrauch. IX, 181 Seiten. 1983.

Vol. 215: Semi-Infinite Programming and Applications. Proceedings, 1981. Edited by A.V. Fiacco and K.O. Kortanek. XI, 322 pages. 1983.

Vol. 216: H.H. Müller, Fiscal Policies in a General Equilibrium Model with Persistent Unemployment. VI, 92 pages. 1983.

Vol. 217: Ch. Grootaert, The Relation Between Final Demand and Income Distribution. XIV, 105 pages. 1983.

Vol. 218: P. van Loon, A Dynamic Theory of the Firm: Production, Finance and Investment. VII, 191 pages. 1983.

Vol. 219: E. van Damme, Refinements of the Nash Equilibrium Concept. VI, 151 pages. 1983.

Vol. 220: M. Aoki, Notes on Economic Time Series Analysis: System Theoretic Perspectives. IX, 249 pages. 1983.

Vol. 221: S. Nakamura, An Inter-Industry Translog Model of Prices and Technical Change for the West German Economy. XIV, 290 pages. 1984.

Vol. 222: P. Meier, Energy Systems Analysis for Developing Countries. VI, 344 pages. 1984.

Vol. 223: W. Trockel, Market Demand. VIII, 205 pages. 1984.

Vol. 224: M. Kiy, Ein disaggregiertes Prognosesystem für die Bundesrepublik Deutschland. XVIII, 276 Seiten. 1984.

Vol. 225: T.R. von Ungern-Sternberg, Zur Analyse von Märkten mit unvollständiger Nachfragerinformation. IX, 125 Seiten. 1984

Vol. 226: Selected Topics in Operations Research and Mathematical Economics. Proceedings, 1983. Edited by G. Hammer and D. Pallaschke. IX, 478 pages. 1984.

Vol. 227: Risk and Capital. Proceedings, 1983. Edited by G. Bamberg and K. Spremann. VII, 306 pages. 1984.

Vol. 228: Nonlinear Models of Fluctuating Growth. Proceedings, 1983. Edited by R.M. Goodwin, M. Krüger and A. Vercelli. XVII, 277 pages. 1984.

Vol. 229: Interactive Decision Analysis. Proceedings, 1983. Edited by M. Grauer and A.P. Wierzbicki. VIII, 269 pages. 1984.

Vol. 230: Macro-Economic Planning with Conflicting Goals. Proceedings, 1982. Edited by M. Despontin, P. Nijkamp and J. Spronk. VI, 297 pages. 1984.

Vol. 231: G.F. Newell, The M/M/∞ Service System with Ranked Servers in Heavy Traffic. XI, 126 pages. 1984.

Vol. 232: L. Bauwens, Bayesian Full Information Analysis of Simultaneous Equation Models Using Integration by Monte Carlo. VI, 114 pages. 1984.

Vol. 233: G. Wagenhals, The World Copper Market. XI, 190 pages. 1984.

Vol. 234: B.C. Eaves, A Course in Triangulations for Solving Equations with Deformations. III, 302 pages. 1984.

Vol. 235: Stochastic Models in Reliability Theory. Proceedings, 1984. Edited by S. Osaki and Y. Hatoyama. VII, 212 pages. 1984.

Vol. 236: G. Gandolfo, P.C. Padoan, A Disequilibrium Model of Real and Financial Accumulation in an Open Economy. VI, 172 pages. 1984.

Vol. 237: Misspecification Analysis. Proceedings, 1983. Edited by T.K. Dijkstra. V, 129 pages. 1984.

Vol. 238: W. Domschke, A. Drexl, Location and Layout Planning. IV, 134 pages. 1985.

Vol. 239: Microeconomic Models of Housing Markets. Edited by K. Stahl. VII, 197 pages. 1985.

Vol. 240: Contributions to Operations Research. Proceedings, 1984. Edited by K. Neumann and D. Pallaschke. V, 190 pages. 1985.

Vol. 241: U. Wittmann, Das Konzept rationaler Preiserwartungen. XI, 310 Seiten. 1985.

Vol. 242: Decision Making with Multiple Objectives. Proceedings, 1984. Edited by Y.Y. Haimes and V. Chankong. XI, 571 pages. 1985.

Vol. 243: Integer Programming and Related Areas. A Classified Bibliography 1981–1984. Edited by R. von Randow. XX, 386 pages. 1985.

Vol. 244: Advances in Equilibrium Theory. Proceedings, 1984. Edited by C.D. Aliprantis, O. Burkinshaw and N.J. Rothman. II, 235 pages. 1985.

Vol. 245: J.E.M. Wilhelm, Arbitrage Theory. VII, 114 pages. 1985.

Vol. 246: P.W. Otter, Dynamic Feature Space Modelling, Filtering and Self-Tuning Control of Stochastic Systems. XIV, 177 pages. 1985.

Vol. 247: Optimization and Discrete Choice in Urban Systems. Proceedings, 1983. Edited by B.G. Hutchinson, P. Nijkamp and M. Batty. VI, 371 pages. 1985.

Vol. 248: Plural Rationality and Interactive Decision Processes. Proceedings, 1984. Edited by M. Grauer, M. Thompson and A.P. Wierzbicki. VI, 354 pages. 1985.

Vol. 249: Spatial Price Equilibrium: Advances in Theory, Computation and Application. Proceedings, 1984. Edited by P. T. Harker. VII, 277 pages. 1985.

Vol. 250: M. Roubens, Ph. Vincke, Preference Modelling. VIII, 94 pages. 1985.

Vol. 251: Input-Output Modeling. Proceedings, 1984. Edited by A. Smyshlyaev. VI, 261 pages. 1985.

Vol. 252: A. Birolini, On the Use of Stochastic Processes in Modeling Reliability Problems. VI, 105 pages. 1985.

Vol. 253: C. Withagen, Economic Theory and International Trade in Natural Exhaustible Resources. VI, 172 pages. 1985.

Vol. 254: S. Müller, Arbitrage Pricing of Contingent Claims. VIII, 151 pages. 1985.

Vol. 255: Nondifferentiable Optimization: Motivations and Applications. Proceedings, 1984. Edited by V.F. Demyanov and D. Pallaschke. VI, 350 pages. 1985.

Vol. 256: Convexity and Duality in Optimization. Proceedings, 1984. Edited by J. Ponstein. V, 142 pages. 1985.

Vol. 257: Dynamics of Macrosystems. Proceedings, 1984. Edited by J.-P. Aubin, D. Saari and K. Sigmund. VI, 280 pages. 1985.

Vol. 258: H. Funke, Eine allgemeine Theorie der Polypol- und Oligopolpreisbildung. III, 237 pages. 1985.

Vol. 259: Infinite Programming. Proceedings, 1984. Edited by E.J. Anderson and A.B. Philpott. XIV, 244 pages. 1985.

Vol. 260: H.-J. Kruse, Degeneracy Graphs and the Neighbourhood Problem. VIII, 128 pages. 1986.

continuation on page 195